Vintage Luggage

BUTTER

Vintage Luggage

Helenka Gulshan

Philip Wilson Publishers

First published in 1998 by
Philip Wilson Publishers Limited
143–149 Great Portland Street
London W1N 5FB

Distributed in the U.S. and Canada by
Antique Collectors' Club
Market Street Industrial Park
Wappingers' Falls
New York 12590

ISBN 0 85667 498 2

Edited by Melissa Larner
Designed by Andrew Shoolbred

Printed and bound in Italy
by Artegrafica

Half title page
Lady's blue morocco dressing-case from Mappin & Webb catalogue, 1934.

Frontispiece
Typical, black leathercloth 'Coracle' brand picnic set.

Title page
From top pigskin sextant-case by Alfred Clark *c.*1935; green-canvas-cased travelling stove by Asprey; unmarked collarbox *c.*1925; cased set of three bottles for toiletries by Allen's *c.*1880; elephant-hide clutch-bag *c.*1930; 'mock-croc' suitcase with cap corners *c.*1930. Front: 'magic' wallet in dark-brown-morocco leather; leather-cased travelling iron; mock-crocodile manicure set; notebook; triangular purse, all *c.*1930.

Opposite
Fibre attaché-cases, including 'The Million' (a popular, low-priced window line).

Contents

M.C.W
W. R.

Author's Acknowledgements

I would particularly like to thank John Gulshan for meticulously gathering information, for his willingness to impart his minute knowledge, as well as for generously allowing the use of photographs.

I would also like to thank Bramah Security Equipment Ltd for supplying the chronological history of Bramah locks, and Mappin & Webb Ltd for kindly granting permission to reproduce illustrations from their catalogue of 1934, and for providing the detailed list of company addresses.

Further recognition, for creative contribution, is due to Ambrose Bennett, Mary Ferdinand, Nick Fowler, Marianne and Michel Franssen, Luciano & Lucinda Gulshan, Alfons, Jon and Patricia Kryszewski, Lynn Martin, Viv Maynard, Jack Meredith, David Page, Richard Whitaker and Tony Wraight.

Opposite
Selection of crocodile luggage. The bag on top, *c.*1900 is marked 'Tiffany, New York'. The golf-bag is *c.*1875, probably of Indian manufacture.

INCORPORATED
FIRST CLASS
PASSENGER'S NAME
CABIN No.
STEAMER
FROM
TO
SAILING DATE
WANTED
IN STATEROOM
KOBÉ
Passenger's Name
LUNNS
PASSENGER
TO
LONDON
CUNARD LINE
CUNARD WHITE STAR
B
FIRST CLASS
PRINTED IN ENGLAND
CUNARD LINE
CUNARD WHITE STAR
S
CABIN CLASS
PRINTED IN ENGLAND
ROYAL MAIL LINE
"ANDE
SAILING:
EMBARKING AT
Berth No.
CRUISE
STATEROOM
BAGGAGE
QUEEN MARY
CHERBOURG
WASHINGTON 25 D.C.
CANADIAN PACIFIC
B
CABIN
ROYAL MAIL LINES
Canadian Pacific
CUNARD WHITE STAR
R
CABIN CLASS
PRINTED IN U.S.A.
WHITE STAR LINE
S
PENINSULAR & ORIENTAL STEAM NAVIGATION COMPY.
R
BAGGAGE
M. M. 698
BAGGAGE
B. I. S. N. Co., Ld.
(INCORPORATED IN ENGLAND.)
CALCUTTA
TO
PENANG
S.S.
Per S.S.
S.A.S.–S.A.R.
NA–TO
TAFELBAAIDOKKE.
TABLE BAY DOCKS.
Vir verskeping.–For Shipment.
H
UNION-CASTLE
BAGGAGE ROOM
NOT
WANTED
ON
VOYAGE
B·I
BAGGAGE ROOM
Shaw Savill Line
THE BLUE FUNNEL LINE LIVERPOOL
KINDLY PLACE THIS LABEL ON THE END OF THE PACKAGE
P. & O.
BAGGAGE ROOM
PASSENGER'S NAME (IN BLOCK LETTERS)
MRS. E. A. HARRIS
SHIP
BERTH or CABIN №
USE THIS LABEL FOR ARTICLES WANTED ON VOYAGE BUT NOT IN CABIN
BERGEN LINE
B
FRED. OLSEN LINE
H
OSLO
WANTED
IN STATEROOM
PORT OF DEPARTURE
SHIP
PORT OF ARRIVAL
FINAL DESTINATION
NEDERLAND LINE ROYAL DUTCH MAIL
G
AMSTERDAM
CABIN
PACIFIC
EMPRESS OF BRITAIN
SEP 28 1935
SOUTHAMPTON
Name
S.S.
Sailing Date
Room
To
WANTED
UNITED NATIONS

1

A Case History

Over the last two centuries, the shape, design and construction of luggage have been greatly influenced by the developing possibilities of travel. As early as the Edwardian age, cumbersome trunks and portmanteaux, which were designed primarily for strength and durability and would literally be 'lugged' to the hold of a ship by minions employed for the purpose, were beginning to evolve towards a more portable alternative. Mid-Victorian luggage was intended almost exclusively for horse-drawn coach and train travel, frequently sent on ahead by the owner and handled solely by servants and porters. By the turn of the century, the advent of the motor car and the rapid spread of personal ownership of these vehicles, coupled with the similar development and commercialisation of air travel from the mid 1920s onwards, brought more radical changes to the concept of modern luggage.

The ever-increasing number of people owning motor cars, liberated from the restrictions of railway schedules, became more independently mobile. As the pace of personal transport accelerated and a newly gained sense of spontaneity and freedom took hold, luggage that could easily be carried to and from the car without the necessity of outside assistance became a requirement even for the prosperous. Manufacturers began to make efforts to make their products suitable for the footloose motorist. The heavy, metal frame, previously deemed an essential element of top-grade cases, was superseded by lighter supportive structures, and slogans promising 'guaranteed lightweight' became the norm. Companies such as Karl Baisch of Stuttgart, J. B. Brooks & Co. of Birmingham and S. Reid Ltd of Fleet Street in London specialised in luggage specifically shaped to fit neatly inside particular vehicles.

Opposite
Selection of luggage labels issued by shipping and cruise companies.

Canvas-covered cabin trunk.

Simple coaching luggage
*c.*1840 (missing outer straps).

Large coaching trunk
*c.*1870.

Compact, four-person wicker tea-basket *c.*1930.

Two U.S. mail-bags in leather and leather-and-canvas, with American 'stagecoach' trunk *c.*1880.

The vast majority of people, however, continued – and continue to this day – to prefer an eclectic selection of classic luggage to accompany them on their travels. That there is still a demand for the highest quality, heavyweight wardrobe trunks and other bulky travelling luggage is demonstrated by the fact that these are still made by the Parisian firm Louis Vuitton (probably the only company with an unbroken tradition of producing these items, which have changed little, if at all, since the Victorian era). But their current product range also encompasses the huge variety of modern developments, such as the soft-sided suit-carrier and the plethora of less structured casual bags that comprise the mainstay of baggage for the mobile modern traveller.

Case presented in September 1927 to Flight Lieut. S.N. Webster R.F.C., pilot of the plane that won the Schneider Trophy that year.

Opposite
Lightweight luggage for early air travel.

The first transatlantic passenger flight took place on 4th June 1927, carrying travellers from New York to Germany. Vulcanised fibre, raffia, wicker, and lightweight metal alloy cases were favoured for use in early commercial air travel, since a combination of lightness and strength was of paramount importance. A traditional leather suitcase, even without contents, could easily tip the scales onto 'excess baggage', and weight tolerances were so slim in the early days of commercial flights that a thoughtless approach to packing could necessitate leaving one's baggage behind.

Created with the same technology as the planes themselves, 'Airport' Lightweight Luggage *c.*1928 was a popular option. Made in England, with registered trade mark, and patent pending, its sales tag proclaimed: 'This case is constructed from light-weight non rusting aero alloy by modern engineering methods. It is of high tensile strength, and guaranteed to give lasting satisfaction. It will not warp, buckle or split and can be cleaned with a damp cloth. It is especially recommended for use in humid or extreme climates.' And indeed, many examples of these cases still survive, often in frequent use.

If Flight Lieutenant S.N. Webster, formerly of the Royal Flying Corps and pilot of the team that took the Schneider Trophy in the famous seaplane race of 1927, favoured such hi-tech luggage, it was not reflected in the choice of commemorative gift presented to him on that occasion. Clearly, a classic cowhide suitcase with cap corners – the traditional alternative – was considered more fitting. Gold-blocked onto the toffee-coloured pigskin interior of the lid is the inscription: 'Presented to Flight Lieut. S.N. Webster R.F.C. by his Fellow Townsmen, with other gifts, to commemorate his great victory for Great Britain in the Schneider Trophy Race at Venice, September 26th 1927'. The surrounding memorabilia of his achievement are still safely stored away inside.

This is just one of many surviving examples of vintage leather luggage, a great number of which bear some clues to their original ownership. These frequently take the form of initials, or a monogram (sometimes made separately of ornate silver and then attached to the case), or occasionally, the full name and address of the original owner – gold-blocked, stamped or painted onto the body of the case. Luggage of lesser quality may still bear a tag, often leather-cased, with details of its owner, and sometimes a record of several previous owners, tucked inside. These telling clues and genealogies by no means detract from

LUGGAGE IN ADVANCE
Spartan

HOTEL MÉTROPOLE BRUXELLES
KÖLN AM RHEIN
HÔTEL MONOPOL PRAGUE
Park-Hotel MANNHEIM
NEW YORKS BEST LOCATED HOTEL
GRAND HÔTEL BRASSEUR LUXEMBOURG
ENTRAL-HOTEL BERLIN
Gegenüber dem Central-Bahnhof Friedrichstrasse
STORA HOTELLET JÖNKÖPING SWEDEN
IMPERIAL HOTEL
KAMPALA UGANDA
HOTEL MONACO & GRAND CANAL VENEZIA
HOTEL WELLINGTON
59.60 DIGUE DE MER OSTENDE
GRAND HÔTEL TOPLICE BLED JUGOSLAVIJA
GRD. HOTEL DES SALINES AU PARC RHEINFELDEN LES BAINS
SUISSE
ARRITZ HOTEL
BEYROUTH
Gasthof-Pension Montfort
HOTEL Washington LUGANO
HOTEL BAHNHOF TERMINUS BRUGG Schweiz SWITZERLAND SVIZZERA SUISSE
BOSCH
HOTEL-RESTAURANT ARNHEM HOLLAND
HOTEL BOSCH
HOTEL D'EUROPE
HOTEL CENTRALBAHNHOF HAMBURG
HOTEL SITGES
San Gaudencio, 5 · Tel. 50
The WORCESTERSHIRE BRINE BATHS HOTEL
DROITWICH SPA
AVIGNON
Hôtel
TERMINUS-GOLF
MONDORF-LES-BAINS
GRAND-DUCHÉ DE LUXEMBOURG
GROSVENOR HOUSE
Park Lane
London, W.
WINDSOR HOTEL
LANCASTER GATE, HYDE PARK, LONDON, W.2
PARKHOTEL AMSTERDAM
LAKES OF KILLARNEY
INTERNATIONAL HOTEL
RESIDENCE PALACE ROMA
IN ROME
MAKE YOUR HOME
GRAND-HÔTEL DU SOLEIL D'OR
HOTEL CONTINENTAL LUCERNE SWITZERLAND
ULM
DONAU
NEUTOR HOSPIZ
(CHRISTLICHES HOSPIZ)
PALACE HÔTEL TURIN
Torino
GRAND HOTEL EUROPE LUCERNE
SWITZERLAND
LUXEMBOURG
PARIS
HOTEL DE LAVAL
HOTEL SONNE INTERLAKEN
DANIELI ROYAL EXCELSIOR VENEZIA
MONTREUX TERRITET
LAC LÉMAN (SUISSE)
HOTEL LONDRES NEW-YORK
Holms HOTEL
HOTEL FAST

the value or charm of a piece. In fact, early inscriptions in neat, copperplate script using Indian ink will often add to the desirability of an item. Modern graffiti, however, untidily scrawled in felt marker or ballpoint pen, and virtually impossible to remove without causing severe damage, will diminish its integrity and originality.

Further traces of the history of these cases – evidence of travel by rail, ship and air – can be found in pasted-on railway tickets and shipping labels. Where the case has travelled by train, separately from its owner, these may state a billing amount in shillings and pence. More often, however, particularly with earlier baggage, they simply bear the point of departure and the destination. Many of the applied paper labels include a space for the date of travel, name, address and destination of the voyager. Steamship labels (of which there are literally hundreds) may give an option as to whether a case was 'Wanted' for the cabin, 'Hold', or 'Not Wanted on Voyage'. A battered old case plastered with such labels will give a surprisingly comprehensive account of the case's itinerary. Shipping labels may refer to one of the great cruise ships, or to the many lines departing for, or returning from, the entry ports that were the gateways of the British Empire. The several distinctive P. & O. (Peninsular and Orient Steam Navigation Co) labels, along with Royal Mail Lines and the later 'Cunard White Star' are often found attached to luggage. The Early 'White Star Line' label, (dating from before the merger with Cunard in the early 1930s), showing a simple white star on a blue or red background, is the rarest and the most sought after of all.

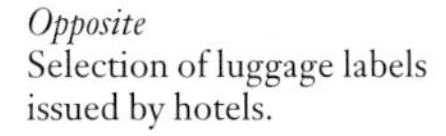

Opposite
Selection of luggage labels issued by hotels.

Early examples from the 1930s of 'Zipp bags' – the type of soft-sided luggage favoured by the modern traveller.

1930s depiction of early air travellers and their luggage undergoing a customs check.

Cases with cap corners
top to bottom
English suitcase *c.*1890 (note unusual locks); suitcase with riveted flat handle; large, well-patinated suitcase with 'fish-tail' handle and metal attachments; large suitcase by John Pound (the side handles render it a 'hybrid' of trunk and suitcase).

Two cased map sets for the early motorist, the large, open one by Phillip Son & Nephew Ltd *c.*1900; silver-mounted purse; cased binoculars marked 'Asprey'.

Throughout the early decades of this century it was fashionable for hotels, when confirming bookings, to issue destination labels to be affixed to luggage. Apart from providing an effective means of advertising, these labels often served a more functional purpose. For example, the proprietor of the Lion Hotel, which opened in the east coast resort of Skegness, Lincolnshire, in 1881, advertised in a guide of 1910 that a porter would meet all trains to carry luggage to the hotel. This was a facility provided by most residential hotels in the town; indeed, those more distant from the railway station supplied omnibuses for visitors and luggage. The presence of a bright, distinctive label must have greatly assisted in the identification of guests. The practice has left us with the legacy of a range of picturesque, imaginative artworks, often depicting the grandeur and appeal of the hotel itself, or local beauty spots and tourist attractions. The variety and abundance of these labels is breathtaking: literally thousands of different examples were produced, originating from little-known guest houses in obscure resort villages in Turkey or the Balkan States, to mighty establishments such as London's Grosvenor House hotel.

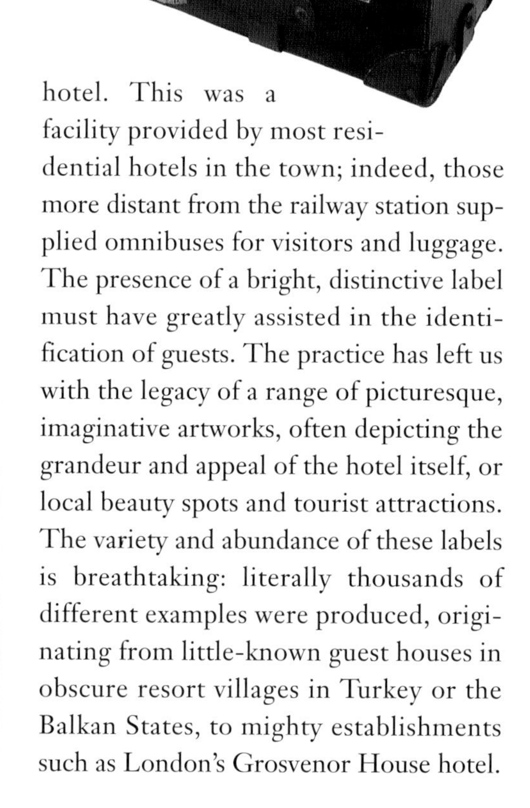

Attached to an old trunk or case, they add further colourful and poignant clues to the personal history of each piece of luggage.

All these labels and methods of identification tell a unique story and can sometimes greatly assist in the assessment of the age of the item in question, or at the very

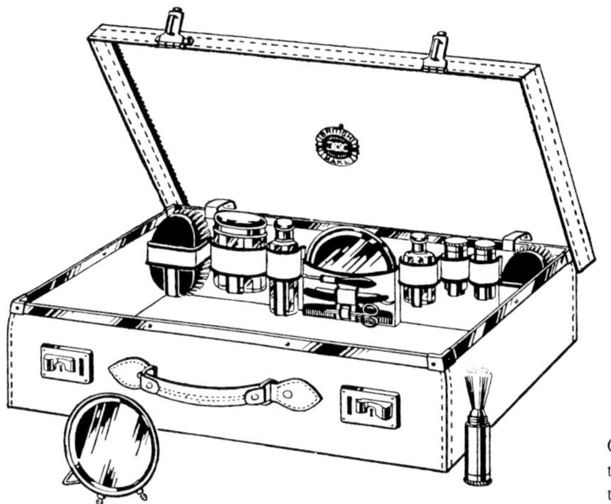

Gentlemen's fitted cases: the 'Popular' *(far left)* and the 'Runabout'.

least, the point at which it was in frequent use. These clues encourage one to carry out further research, to arrive at explanations and theories, and, in the process, to learn about a broader past – to delve into history, unearthing fascinating details and echoes of the lives of our recent ancestors.

2

A Suitable Case

Given the sheer physical weight of antique luggage, and the outmoded function of some of the quainter pieces, many people who collect it provide it with new uses: collarboxes are used to keep cosmetics or medication tidy, hatboxes for photographs and love letters, trunks to store magazines or spare bedding. Others may not demand that it fulfil any practical role at all, merely enjoying having it around, to appreciate and treasure as a relic emblematic of our heritage.

Examples of the earlier, far less portable trunks are rarely utilised for travel in the present day, but are sought after nonetheless for use as low tables and for storage. When their beautifully patinated leather has been cleaned and polished to a gentle glow, their soft, autumnal colours accented by toning studs and fittings of burnished yellow brass, they lend a mellow character to designer rooms and to the shop-window displays of clothes stores. Such well-travelled and romantic articles convey a sense of quality that has stood the test of time: many of the 'theme' restaurants and bars, whose popularity owes much to nostalgia, use these items as 'props', along with sporting artefacts and books with decorative bindings to create an ambience of warmth, security and substance.

Recent costume dramas of stage and screen have also led in a subtle way to an increased appreciation of the beautifully crafted cases of yesteryear. Even the bulky items produced in the mid-to-late 19th century, deemed until fairly lately to be obsolete except for their curiosity value, appear in a very different light after one has watched the adventures of Indiana Jones with his heavy, battered kit bag, which survives many strenuous adventures. Such atmospheric productions as *Out of Africa* or *The Remains of the Day* charmingly illustrate, by the use of traditional luggage, amongst many other artefacts, the atmosphere and flavour of a bygone age.

Opposite
Eight classic, leather suitcases. Note the wide variations of colour, handle fixings and lock design. The right-hand case has Garstin's 'Klip-it locks'.

The 'Identity' trunk, offered in a choice of colour combinations in the early 1930s.

American 'Innovation' brand wardrobe-trunk ('Pat'd May 17 1898 – Jan 30 1900'), which hinges open vertically rather than horizontally.

Opposite
Louis Vuitton 'Ideal' trunk – rare and desirable in any fabric, but even more unusual in leather. The lid compartments were useful for the transportation of golf-clubs, walking-canes and umbrellas.

Many small suitcases, hatboxes, vanity- and writing-cases, however, are still sufficiently durable to be employed for their original purpose, and if discrimination is exercised it is possible to build a collection of vintage luggage suitable for everyday use. This may require time and dedication, for trouble must be taken in maintaining and repairing it, but the resulting pleasures of travelling with classic leather suitcases make this effort well worthwhile to an increasing number of serious collectors.

The strong appeal of collecting antique leather goods can be attributed to many specific factors. The enticing trail of clues for the historical detective has already been discussed, and this is perhaps part of a more general fascination with the past experienced by all collectors. While the sight of a beaten-up kit bag may leave one person completely cold, they may be totally enthralled by a stack of early *Beano* comics or an original Barbie doll, an art deco lamp, or an example of Martinware. Nostalgia plays a large part in the aura surrounding old, collectable things.

However, the allure of works of art or beautifully crafted goods designed for occasional use, such as a Fabergé egg or a Lalique glass item, is vastly different from the attraction of items made of seemingly less permanent, organic substances like leather, which were fashioned for immediate, functional use. It is doubtful that the manufacturer of a bag in the 1880s, however lovingly he made it, would imagine that it might survive, still cherished and functional, to the present day. Practical objects are likely to be much used, but appropriate design for function, along with the highest standards of workmanship, have enabled much antique luggage to

'The Auto' – an 'Everwear' brand wardrobe-trunk, retailed at a low price in 1933.

outlive by far the durability envisioned by its maker.

Many people find the intrinsic transience of leather as a substance unsettling. It is, after all, the skin of an animal and, unlike glass or metal, organic in origin, albeit preserved by the process of tanning. Leather is potentially evanescent – it seems miraculous when an ancient suitcase outlasts its original owner. But this is the very reason why an aura of mystery and romance surrounds even a poor-to-medium-quality trunk or suitcase from the earlier decades of this century. Its hard edges have been softened and smoothed by time and travel, it has mellowed, and could almost be said to have taken on a life of its own. It is the ephemeral aspect of antique leather that lends it character and delight. The fact that someone has obviously loved and enjoyed it is an intangible, but nonetheless potent force, and the idea that, with care and attention, one may continue to love and enjoy it adds to the pleasure of owning it. And it is possible to maintain and conserve it almost indefinitely. For proof of this, one only has to recall the perfectly preserved leather sandal, well over a thousand years old, that was dug out of an Irish peat bog. Leather suffers most in extreme or direct dry heat, and a collection kept in a centrally heated environment must be regularly replenished with moisture or it will desiccate and crumble to dust if handled. Various preparations that will prevent leather from drying out are readily available.

EVERWEAR REGISTERED BRITISH MAKE TRADE MARK WARDROBE

Of course, when buying antique luggage, it is crucial

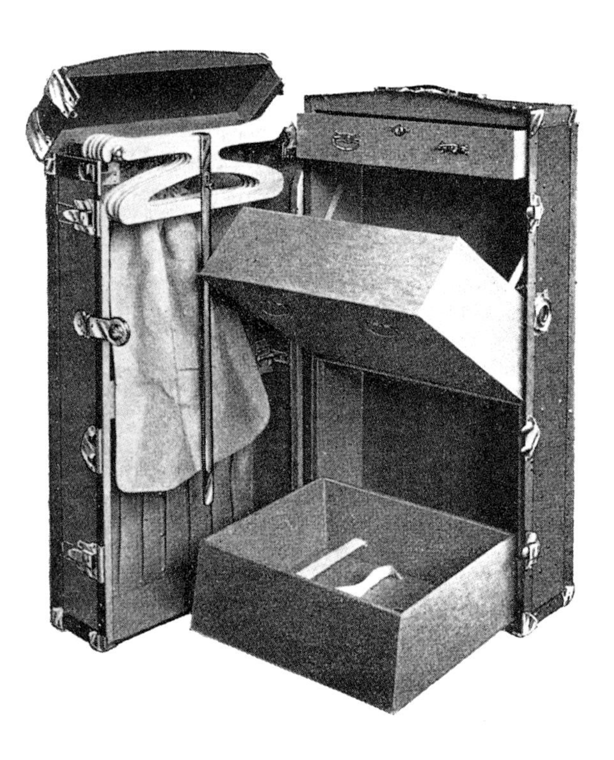

Another 'Everwear' brand wardrobe-trunk, offered in a wide variety of colour combinations.

that choices should be made carefully. The most important factors to consider when appraising a piece are quality of materials and construction, balanced against physical condition. Damage and decay of leather can be arrested, but never reversed. Preservation is an easy task, but damage already apparent must be addressed separately. Inspect the back seam of a lidded case with great care. If it is torn or worn, are there any rivets or studs that will make it difficult to facilitate a sympathetic strip- or patch-style repair? If the structural seams are damaged, is it because the leather has rotted away at the edges, or perhaps merely a matter of getting a little re-stitching done with matching linen thread, re-using the original and still-sound stitch-holes?

There are many pointers to quality, and for a novice in the area of antique and vintage luggage it is probably wisest to look for a maker's mark (see Chapter 5). Any manufacturer truly proud of their merchandise tended to take the opportunity to advertise its provenance, and this is just as true today with designer and branded goods. These marks are usually stamped on the front rim of the base, in the centre under the lid. They may also be found on the interior lid, either stamped, or on an applied label. A select handful of makers, including Finnigans Ltd, of London, W. Insall & Sons of Bristol, and Brachers (Bristol & Cardiff), impressed their name directly into the hide in a small oval cartouche (or scroll-like frame) at the centre of both lateral edges of the exterior lid. Gladstone and kit bags are usually stamped on the inner-top rim of the frame, or on the lining, and occasionally on the base of the bag.

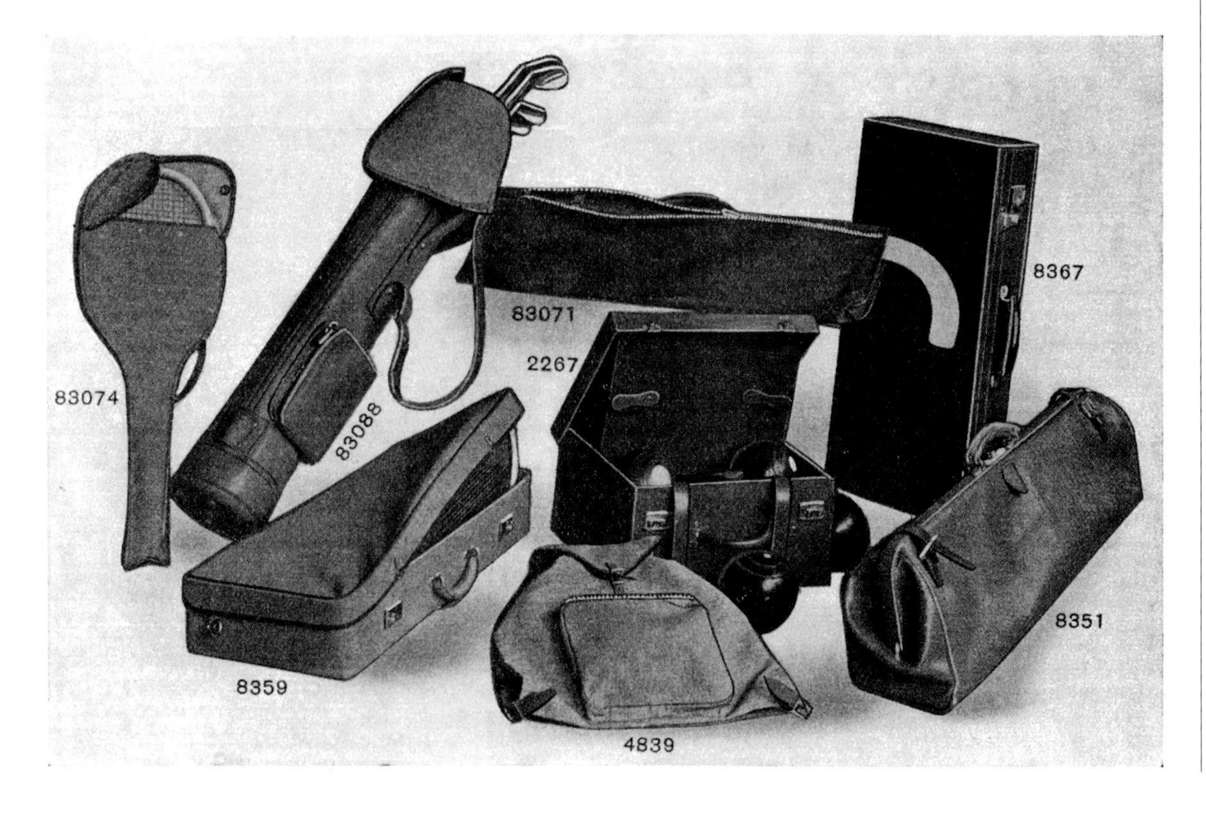

Cricket, golf and hockey bags, tennis and bowls-cases.

Selection of kit bags.

Generally speaking, the only makers of fine English luggage to have consistently stamped brass and nickel-on-brass lock fittings with their name, in addition to marks on the body of the case, are Finnigans, F. Lansdowne (London) and W. Barrett & Sons (London). If the lock is stamped 'Bramah' or 'Chubb' (the latter, on earlier, rare items, accompanied by a lively engraving of the fish of the same name), this indicates that the fittings, but not necessarily the case itself, were produced by that maker, unless further marks are present on the body of the case. John Pound & Co. (London) routinely stamped 'JP' on locks. Since the late 19th century, when they patented their own lock, Louis Vuitton have stamped their locks with both their name and a lock number (separate from the serial number

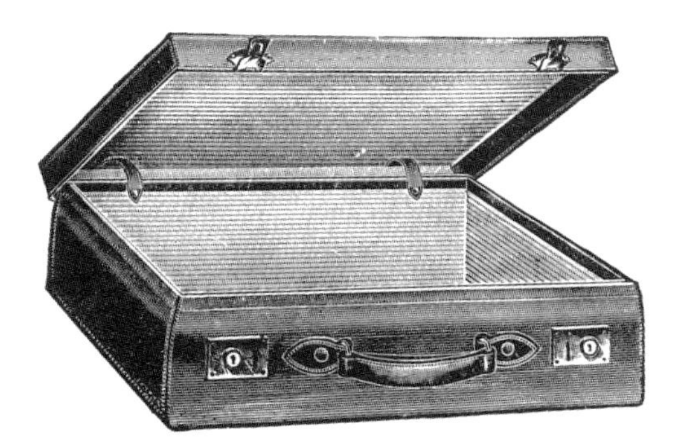

applied to the interior). However, it is fairly commonplace to stumble upon an earlier, caramel-and-toffee striped Louis Vuitton with unmarked, squarer, pre-patent locks. This is unlikely to be entirely original if the locks are marked in any way.

Many locks on English leather cases and trunks are marked 'SECURE LEVER', 'SINGLE LEVER', 'TWO LEVER', 'THREE LEVER', or even 'FOUR LEVER'. This is not a maker's name, merely a description of the type of lock. There are also numerous examples that are stamped 'ENGLISH LEVER', 'ENGLISH MAKE', 'MADE IN ENGLAND', 'BRITISH MAKE', 'MADE IN BRITAIN' or, on occasion 'LONDON MADE'. These are all pointers to that little extra touch of pride, and still herald a degree of quality in the absence of an actual maker's name.

The colour of the suitcase is also important. If it is a blandly even, flat-looking, orangey or plain-chocolate brown, or glossy navy blue, with a sprayed or painted finish it is unlikely to be really desirable, whatever its age. If you will enjoy using it, however, it may still be a worthwhile purchase and represent better value than a similar brand-new suitcase. Nevertheless, one should bear in mind that leather goods with a very flat, almost plastic, applied finish cannot truly be termed 'vintage'. A really collectable item is usually made of hide treated in a more natural way, which has enabled the skin to experience life – as your own skin does. It will have acquired a mellow patination, that depth of surface so elusive of description.

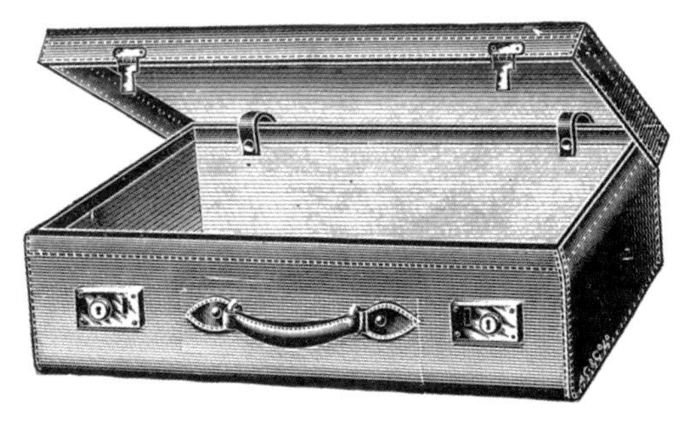

Medium-quality cowhide suitcases of the 1930s.

An expert in fine antique furniture will describe in no uncertain terms how fragile is the valuable element inherent in antique wooden furniture, and how easily genuine patination may be carelessly destroyed. The patination of a fine piece of 200-year-old mahogany – that glorious, soft and shimmering glow of antiquity, long use and loving care that lifts it out of the ordinary and enables it to speak to us of the past – has the superficial thickness of two sheets of paper. Less than a quarter of a millimetre of physical depth, which can easily be sanded away by an over-zealous restorer, stands between a fine old original and a reproduction. Leather is akin to wood,

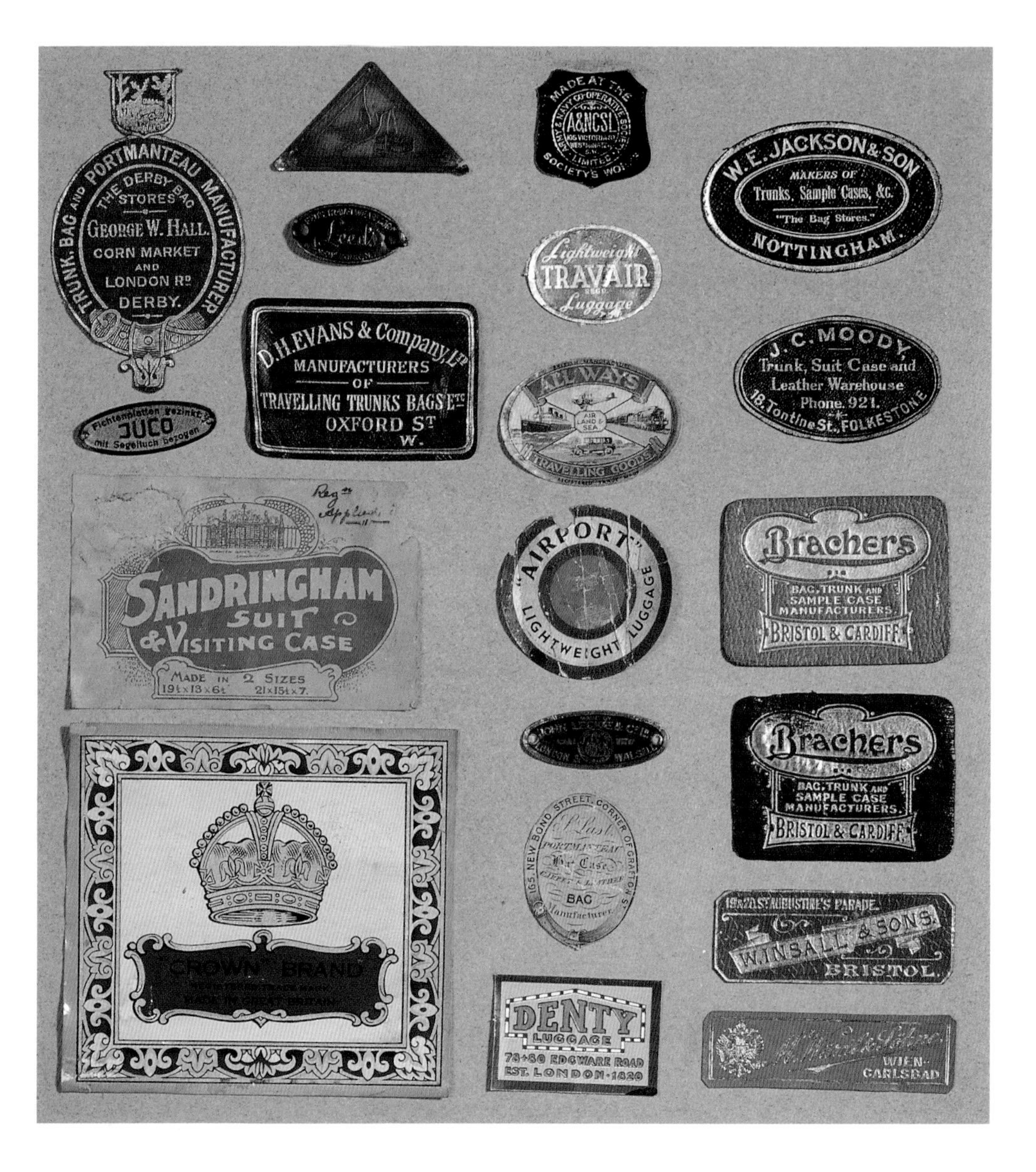

Unusual applied labels.

Opposite, top to bottom
Top-hat box *c.*1880; leather quiver with hand-made brass lock *c.*1850; Louis Vuitton leather wardrobe-trunk, closed and on its side, in fair condition (compare with illustration on p.80); Louis Vuitton courier trunk in fair condition.

another organic material, but is in general far less dense, and thus acquires patination at approximately three to four times the rate of wood.

When a case has been kept in its original, fitted 'foul-weather' cover for many years – often with a separate cover for the handle – it may have survived in an almost 'as-new' condition (see illustration on p.29). The cover itself, generally of canvas and frequently trimmed with hide to match the case itself, will have afforded great protection through the years and it is likely that it will be battered and torn. Every bit of damage or wear and tear evident on this overcoat would otherwise have been suffered by the case itself, and the exceptional state of preservation of cases protected in this way bears witness to the fact that the cover was a worthwhile feature. An unusually well-kept item, whilst lacking the obvious romantic appeal of the more

Compare *top to bottom*
Low-grade cardboard suitcase, brass locks, *c.*1920; low-grade compressed-leather suitcase, brass locks, *c.*1920; typical leather case made in the Far East *c.*1930; standard, chocolate-brown leather suitcase with no maker's name; unmarked but highly desirable case (note colour, patination, brass locks, unusual proportions, original key).

careworn piece, is of great academic interest, since it will convey a striking impression of the extremely high standard of craftsmanship and quality formerly expected as the norm – even the highest quality modern piece will not bear comparison.

If a manufacturer went to the trouble of custom-making a separate canvas cover for a case, it is probable that the case was made of crocodile skin, or the finest grade cowhide, and that it was hand-made (i.e. hand-stitched or 'hand-closed') rather than machine-made. While the merits of the mass produced should not be dismissed out of hand (there are in fact many interesting, useful, and eminently attractive mass-produced items that have now become as rare as specially commissioned trunks, hatboxes and dressing-cases), hand-stitched luggage has an appeal similar to that of couture clothing. The stitches are fewer to the inch, and less even in an almost imperceptible way, and there is an individuality communicated via the tactile element that cannot be expressed verbally. When a skilled craftsman has worked long and hard on such a piece it is instantly recognisable as a work of art, rather than a mass-produced product. The difference is infinitesimal yet immense.

It is also likely that if a piece has been made with great care, its fittings will be of an equal standard. Solid, cast-brass locks and catches signal quality. But when judging a piece, one must view it in context. A puzzling example of this can be found in a particular group of small, bullet-shaped 'doctor's bags', in which there was a vast disparity between the quality of the hide and workmanship and their metal fittings, which were of tin, rather than brass. When another of these later turned up, this time with the addition of an inscribed, white-metal plaque dated 1915, the mystery was solved: no brass was available for frivolous use in wartime.

Generally speaking, however, if the metal is a thin, silver-coloured, stamped-out tin, probably rust-flecked, it is a negative pointer. If it is of silver colour but heavy quality and obviously cast metal, it is likely to be nickel, or even (on later locks) chrome over brass and will be of extra-fine quality. Sadly, over the past twenty years, there has been a tendency amongst specialists to strip such nickel or chrome plating down to the underlying brass in order to make the quality more obviously apparent to the casual customer. Therefore, many locks that would, and should, have retained their original coating have been sacrificed to fashion and easy commercial gain. Ironically, items on which the chrome, nickel, or even in some instances, gold plating is in good condition are therefore becoming far rarer, and these original, untouched pieces will be much more valuable than the ones that have been modified by dealers eager to maximise profits.

If the metal is a dull, smooth brown – not the flaking, more granular orange-brown of rust, present on many tin locks – or if it is encrusted with green verdigris, it is most probably brass, and may be

Two crocodile cases with foul-weather covers.

polished to restore the shine. If the metal is shiny yellow, on relatively heavyweight cast metal, it is either recently burnished brass or gold-plated brass. Fittings of stamped-out thinner metal, assembled from separate pieces, and apparently having the attributes of brass, generally will be of that metal, but very occasionally lower-grade white-metal locks were coated with either gold or brass plating. Any attempt to polish these lower-grade fittings leads to the coating coming away to reveal the white metal underneath.

Metal fittings made of solid silver are very rare indeed. A few examples exist, however, and are recognisably different. Almost as rare, but by no means as desirable, are cast locks made of white-metal alloy, probably 'gun-metal'. One might also come across lightweight, stamped-out alloy locks, produced from the early 1930s, which have the appearance of aluminium and are referred to as such. The use of any white metal other than silver tends to reduce the desirability of an item of baggage. An exception would be an early, hand-forged (and therefore unique) blacksmith-made piece. Such craftsmanship is, as ever, a distinctive and attractive addition, and renders the materials employed largely irrelevant.

Top to bottom
Rectangular top-hat box by Finnigans *c.*1910; classic square top-hat box *c.*1880; front-handle, hide, gentleman's hatbox (note recent saddler-replaced handle and straps); square canvas hatbox with handle on top (*right*).

3

Making the Case

The majority of hide luggage, even of the most expensive grade, was made over a wooden or cardboard, and later, possibly wood-ply or compressed-fibre, foundation. Only items that were required to be exceptionally durable were made on a solid oak or mahogany foundation, notably gun-boxes, cartridge magazines (and the occasional writing-case, usually on a pine foundation). The term for a case with such an interior construction is 'oak-lined' or 'mahogany-lined', and the material used as a basis was left plainly visible around the inside rim of the case. Two extremely desirable makers are the gun manufacturers, Holland & Holland Ltd (London) and James Purdey & Sons Ltd (London). Both companies are still trading, Holland & Holland stocking a range of antique luggage, and Purdey producing a classic cartridge-case made to a traditional design and standard. Aficionados of quality have been known to remove the wooden dividers from these cases to enable them to function as beautiful, if somewhat heavy, attaché-cases. Other gunmakers producing exceptionally attractive cases were W. Powell & Sons (Birmingham) and William Evans (London).

Wooden-hooped trunks, now widely referred to as 'wood-banded', are normally covered with brown or green canvas (and, less often, with leather or vellum). They are surrounded with protective slats or hoops of bentwood, on a flax-fibre or plywood base. Another common type of trunk is the large, domed black or green canvas tropical trunk, trimmed with brown leather and constructed over a wicker foundation. Black, green or brown, and occasionally cream, canvas cases, hatboxes and trunks trimmed with tan leather and often generously embellished with large, decorative brass studs are worthy of collecting, especially the earlier items. However, standard green canvas, tan-leather-trimmed cases, later of

Medium-quality fibre trunk.

beige canvas trimmed with chocolate-coloured plastic, were produced by the thousand from 1945 for demobilised British soldiers and, whilst of admirable quality and construction, are too commonplace to be interesting to the serious collector. The subtle distinctions between the earlier, custom-made individual cases designed for travel to the tropics, and the latter, mass-produced product become obvious on close inspection.

The most consistently excellent luggage was made of cowhide, treated and manufactured to many grades and thicknesses for a wide variety of uses. Luggage made from bridle- or harness-quality hide wears extremely well. 'Connolly' hide – that bastion of English heritage – has been employed in the production of motor-car seats from the time they were first invented through to the present day and is also used in the manufacture of much of the finest luggage. It is still produced with a painstaking attention to detail rare in the present day, combining both traditional and more modern methods and equipment. The accumulated fund of knowledge and expertise of Connolly Leather Ltd (London) is frequently called upon to recreate and colour-match hide from old samples so that the upholstery of a graceful old vehicle may be exactly restored to its original state.

No doubt an entire book could be written about the different varieties of cowhide and the various methods with which it was treated to prevent decay and to render it fit for its intended use. However, since we are mainly concerned with the end-product as luggage, a brief outline should suffice to convey some idea of the complexities involved in the preparation of leather.

The treatment, which is called 'tanning', is an ancient art. Leather can be made in dozens of different forms, from the extremely tough and stiff material needed for the soles of boots, to very thin and pliable glove leather. Firstly, the skins were removed without damaging them (a very difficult task in itself); at this point they were termed 'green'. If it were not possible to send the skins for immediate tanning, they were either hung to dry out thoroughly, or salted – the object being to prevent the growth of microbes, which cause decay. Skins thus treated were termed 'dry'.

High-grade fibre trunk and round-top trunk.

Opposite
Round-top trunk.

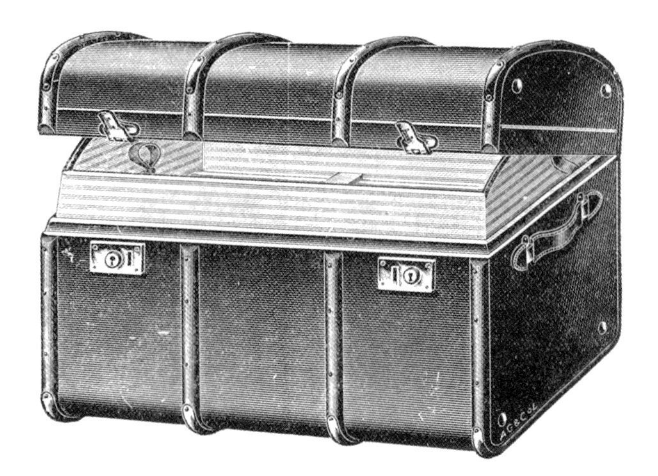

All skin is made up of three layers: the outside thin layer, or 'epidermis', the interior fatty or fleshy layer, and the thicker middle layer known as the 'derma' or true skin, which is the material from which leather is made. To prepare skins for tanning, they were subjected to a variety of complicated processes, according to the kind of skin and the condition in which they arrived at the tannery – whether 'green' or 'dry'. All these processes consisted mainly of soaking the skin in chemical baths, then removing the epidermis and any hair from one side (known as the grain-side) and the fatty flesh from the other, leaving the middle layer of skin ready to tan. At this stage, the skins were referred to as 'pelts'.

The three main methods of tanning were by animal, vegetable or mineral means: using certain oils (generally obtained from fish), tannic acid (from tree bark), or mineral products (usually alum or a chromium salt). The oldest method, and that in most common use at the time under consideration, was the vegetable method. The tannic acid, or tannin, most frequently employed in English tanneries came from the bark of oak and hemlock, or from chestnut wood. Russia leather, a strong, pliable type, was tanned initially with willow bark, and thereafter treated with birch bark. Levant-morocco leather, made of goat or sheep skin, was treated with tannin obtained from the sumac tree. Larch was also used, along with many trees and plants from other lands including the acorn cup of the Turkish or Greek oak, known as 'valonia' or 'vallonia'; Australian mimosa; Indian *myrolobans*; South American *quebracho* and *divi-divi*; Malayan *gambier*; and galls (growths caused by an insect that punctures the bark in order to deposit eggs) from oak trees everywhere. Synthetic tannin was also being produced by the late 1920s, possibly earlier. The skins were immersed in vats of water containing tannin, a solution known as 'drench', moved daily to a stronger solution, stirred, taken out periodically to drain, and finally left to soak in the strongest solution. The process took approximately three months, and was utilised especially for heavy leathers.

In mineral tanning, the use of alum was the oldest process, but towards the end of the 19th century the employment of

Cases made by Nathaniel Benjamin for Mr J. Gordon-Duff.

chromium became fairly widespread. Skins to be tanned were first 'pickled' in a chemical bath, and then treated with a solution of chromium salts. The salt remaining in the fibres of the leather caused a chemical reaction that made the resultant leather more able to endure heat and moisture than that treated by the vegetable method. The process, which took about a month, was used mainly for lighter leathers, such as glacé kid (a thin, highly glossed fancy leather prized for its superbly delicate decorative effect). Oil tanning was also an ancient process, usually involving cod-liver oil. In this extremely complex method the idea was to remove all the moisture from the leather and replace it with oil, but most of the oil also disappeared in its turn from the leather. This process was used for very soft glove and chamois leathers.

After tanning, the leather was dried again, which left it coarse and stiff, necessitating a further process to make it more flexible. This action, called 'currying', involved rubbing grease into the leather, after which it was ready to be dyed as necessary. Depending on the type of leather it was destined to become, it underwent further treatment. Harness or bridle hide, for example, had to be made tough and firm, and was therefore pressurised by mechanised rolling or hammering. For soft and thin leather, the hides had to be split into finer layers. To obtain a highly polished finish, the leather was treated with a certain mixture and then 'glazed' by a glass (somctimcs agatc) cylinder called a 'glassing jack', which rapidly rubbed it to a fine finish, under high pressure. At this stage, the leather was ready to undergo a decorative pressed texture if so desired.

In terms of antique luggage, the general rule is that the thicker the leather, the better the quality of the finished item. However, this has to be balanced with its function. A silk-velvet lined, finely crafted glacé-kid jewellery box, for example, could not be fashioned of cumbersome bridle hide. If a case is brown in hue, with a smooth finish, and not unduly delicate, then it is most likely to be made of cowhide. It is probable that the words 'Guaranteed Cowhide' or 'Warranted' or 'Genuine Cowhide' will be stamped onto the leather or on a leather patch somewhere on the lining of the case.

The most unusual process developed for the manufacture of hide accoutrements is that of Norfolk Hide (registered under Patent Number 134403, but not always

Opposite
Leather wardrobe-trunk by Louis Vuitton.

marked as such), which results in an entirely distinctive moulded effect. At first sight, an item of Norfolk Hide is a puzzling thing, recalling an early plastic such as Bakelite. The process renders the leather so rigid that if you drum your fingernails on it, it feels a little like vulcanised fibre – but it is far thicker, and has more visual and tactile density and depth of patination than such a material could possibly possess. It also retains a strangely animal aura. The parts are all of a piece, with no apparent joins. Although rigid, it is not brittle like plastic, but the fact that there are no structural seams or stitching belies its status as leather. Possibly, on the back seam, where metal hinges have not been used to fix the separately constructed, unstitched base and lid together, a strip of more flexible bridle hide will be sewn to top and bottom through drilled holes. The handle will be of bridle hide. Circular objects made of Norfolk Hide, such as collarboxes, display a faintly discernible series of concentric grooves, less pronounced, but similar in appearance, to those on a 78 r.p.m. record. These products are amongst the most exciting curiosities that one can happen upon in the field of antique luggage. Norfolk Hide is one of the very best-kept secrets harboured by collectors and dealers in fine luggage, and is extremely valuable.

If a case is obviously made of some type of animal skin, but is textured, it is probably cowhide. The most common variety of textured or 'oak-grain' leather is similar to the type that Louis Vuitton has re-launched in recent years as 'Epi' leather. This is embossed with a design of slightly irregular, not quite parallel vertical lines, a little like an elongated tree-bark effect. It is less prone to marking and bruising than smooth leather, and most Gladstone bags, Wellington bags, kit bags and doctor's bags were made of this textured leather – hence its earlier name of 'bag hide'.

Gladstone bags were named after the 19th-century British Prime Minister, William Ewart Gladstone, although there is debate as to whether he ever actually used

Opposite
Selection of Norfolk Hide. Note construction and back seam. The bottom case is unusually large for an item of Norfolk Hide.

Leather attaché-cases.

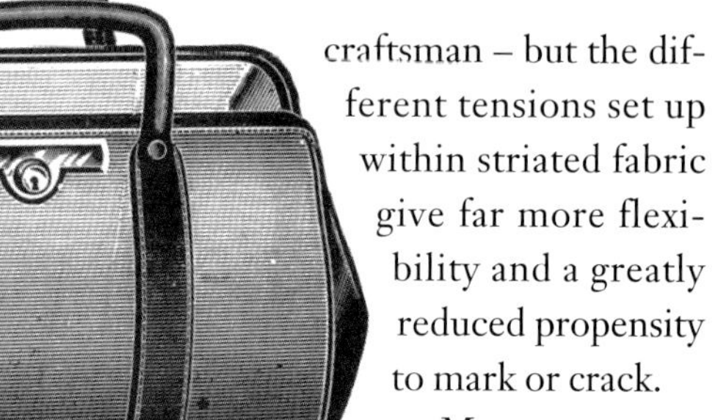

this type of bag. Its ingenuity lay in its design: the leather was fitted over a metal frame that split in half to enable both portions to lie flat when the bag was opened (see illustration on p.107). There are various styles of metal-framed, all-purpose bags and these are often mistakenly identified either as Gladstone bags, or doctors' bags. The illustration at the bottom of this page, for example, is of a brief bag, not a doctor's bag, although from the outside, the two are identical (and see opposite for illustrations of kit bags, commonly mistaken for Gladstone bags). A row of brass buttons around the lining of a doctor's bag provides hard evidence that it genuinely functioned as such. These secured a secondary lining made of white fabric, easily changed and boiled to ensure a sterile environment for medical supplies and equipment. The Wellington was a term coined by Army & Navy Co-operative Stores Limited to describe a particularly capacious kit bag.

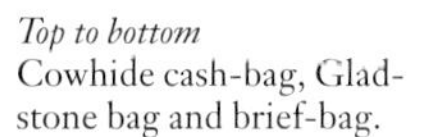

Top to bottom
Cowhide cash-bag, Gladstone bag and brief-bag.

These were rarely, if ever, made from smooth hide because, like a garment, they were sewn inside-out, especially where piping was used. Smooth leather will crease severely if it is turned right-side-out after sewing – however skilled the craftsman – but the different tensions set up within striated fabric give far more flexibility and a greatly reduced propensity to mark or crack.

Most morocco leather – a finely rolled-out version made from sheep or goat hides tanned with sumac – was impressed with a similar pattern. Morocco leather was used mainly for accessorics within cases and for lining, as well as for the exteriors of dainty writing-cases and jewellery-boxes. There are many variations on the theme: one type of morocco leather is stamped with a fine, criss-cross pattern, which was called 'X grain' by Waring & Gillow Ltd (see dressing-case illustrated on p.131), and is very similar to the design that Louis Vuitton now refers to as 'Taïga' leather. This was mostly used in the interiors of luggage. Another, termed 'shagreen' leather, imitates the tiny, round scales of the 'shark skin' of that name, which itself was actually made from Japanese sting-ray skin.

There has long been a market for leather goods stamped and pressed to imitate exotic hides, crocodile in particular. Some modern imitations are extremely difficult to distinguish from the real thing, but it is possible to develop a 'feel' for genuine skins in vintage items by handling

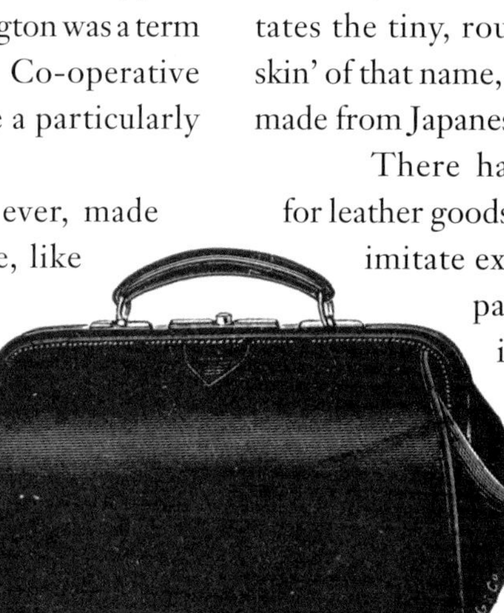

as many examples as possible. (see Chapter 6 for information on the rules and regulations pertaining to genuine exotic hides). The main element to consider is the regularity of the pattern – if it is very even and exactly repeated, it is not going to be genuine crocodile. Also, look for evidence of separate scales – a pressed or stamped leather item is just that: a design embossed onto a smooth hide. A genuine piece of crocodile or alligator skin, however smooth and fine, will reveal in some places evidence that the surface of the fabric is actually composed of differently shaped, distinct and individual scales. The smaller scales may show a little pit or dot, like a pore, however faint. This is absent from pressed leather of the period.

Of the many different species of crocodile, the Common crocodile of West Africa, the Nile crocodile and the American alligator appear to have been the most widely hunted for their skins and, although it is difficult to distinguish with certainty from which species the hide was obtained, the size of the scales and place of manufacture may enable a fairly accurate guess.

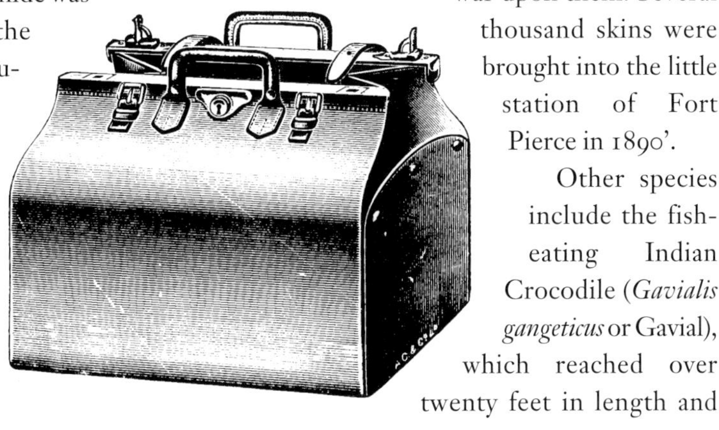

The Common Crocodile (*Crocodilus cataphractus*) ranged from Senegal to the Congo. It was distinguished by its slender snout and its refusal to enter brackish water. The Nile Crocodile (*Crocodilus niloticus*) was also essentially African, distributed across Madagascar, Senegal, the Cape and Egypt, but even in the late 19th century it was described in *The Cambridge Natural History 1909* as 'practically exterminated' in Lower Egypt. It grew to no more than fifteen feet in length. There were two species of alligator: *Alligator sinensis*, not named until 1879, which was found around the Yangtse-Kiang in China, grew to only about six feet long; *Alligator mississippiensis* was widespread near the mouths of rivers from the Rio Grande up to the river Neuss in North Carolina. Most females grew to a maximum of seven feet long, but the males could reach over twelve. A late 19th-century account of the American alligator given by S.F. Clarke stated: 'The alligators are rapidly diminishing in numbers under the stimulus of the high prices offered to the hunters for their hides. Both Whites and Indians make increasing war upon them. Several thousand skins were brought into the little station of Fort Pierce in 1890'.

Other species include the fish-eating Indian Crocodile (*Gavialis gangeticus* or Gavial), which reached over twenty feet in length and

Classic cowhide kit bags.

Opposite
Selection of crocodile luggage. The bag on top, *c.*1900, is marked 'Tiffany, New York, *c.*1900'. The golf-bag is *c.*1875, probably of Indian manufacture.

could be found in the basins of the Ganges, Brahmaputra and Indus rivers; the Marsh Crocodile (*Crocodilus palustris*), held sacred by the Hindus, which inhabited the rivers, ponds and marshes of India, Ceylon, Burma, Malacca and most of the Malaysian islands; *Crocodilus porosus*, which often grew to twenty feet in length and was widely distributed across the Gulf of Bengal and the Malay Archipelago; the smaller *Crocodilus johnstoni* of Northern Australia and Queensland; *Crocodilus intermedius* of the Orinoco, which grew to thirteen feet. *Crocodilus americanus*, distinguished by the bump on its snout, was the only crocodile to inhabit the islands of the West Indies, and was also found in Florida, through warmer areas of central America into Colombia, Ecuador and Venezuela.

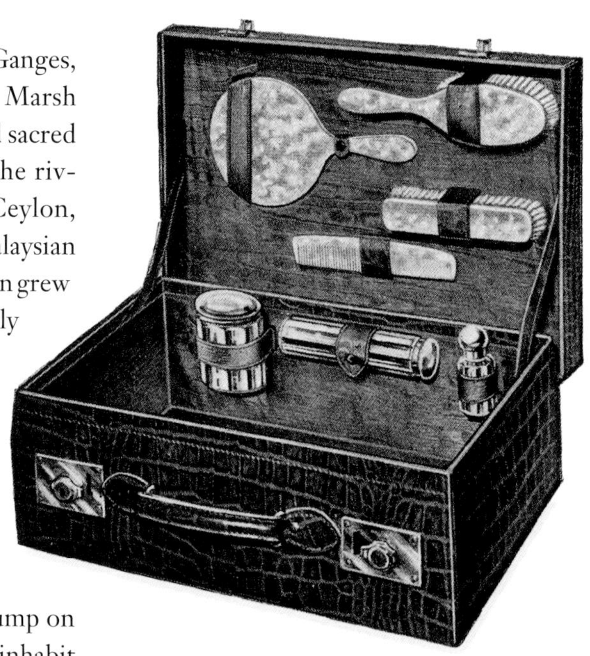

'Splendid Value in Ladies' Fitted Suit Cases': 'Maureen' *(right)*, a model in mock-crocodile, and the cheaper 'Psyche', both on sale in the early 1930s.

As late as the mid 1930s, a firm by the name of E.J. Cooke Ltd (Manchester) was advertising 'WALRUS LEATHER IN HIDES AND STRIPS'. Along with elephant hide, walrus was perhaps the most durable of all leathers. It is very rare that one will find either made into luggage – only a handful of examples still exist. This is possibly because they were not very popular. On a walrus-skin, Gladstone-type bag, the hide looks much as it would do on the animal itself: it is extremely thick, strong and not terribly pleasing aesthetically. But that very toughness made it eminently suitable for industrial use. Elephant hide is similar in appearance to walrus, but tends to be more supple. However, it can be difficult to differentiate between the two.

Ostrich skin, on the other hand, is easy to recognise. It is dotted with large, uneven pores ('pearls') and at first glance appears to be a stylised, floral pattern, stamped in relief, blossoming across the leather in a figuration resembling a sepia depiction of a displayed peacock's tail. Many handbags are available; writing-cases are rare, and much more desirable.

M.C.W.
W. R.

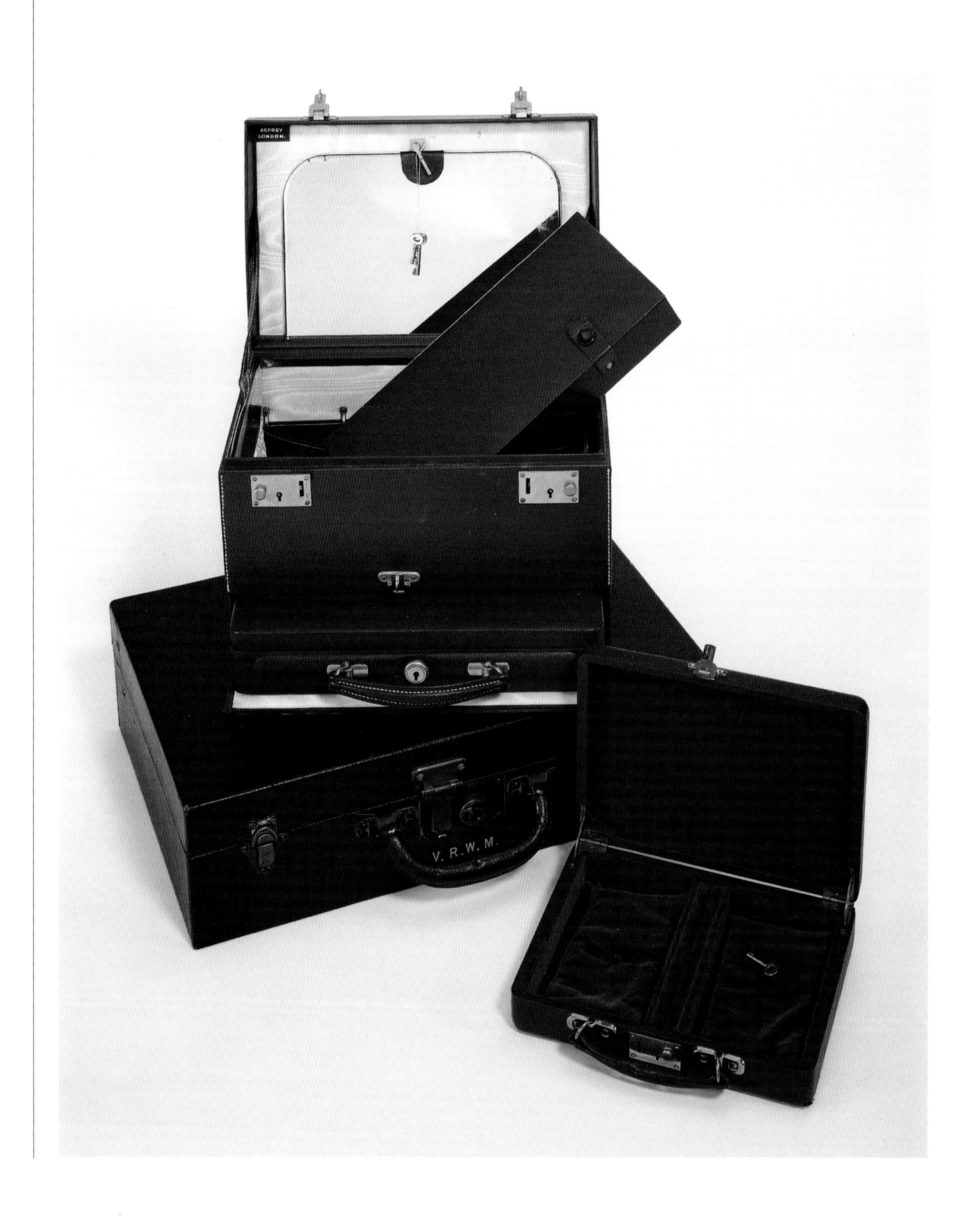
ASPREY
LONDON.
V. R. W. M.

Lizard and snakeskin were occasionally used as linings and a few small suitcases and round hatboxes were also made from these skins, but generally their use was limited to smaller items such as shoes, purses, clutch-bags and handbags, mainly from the period between the wars.

This era was the heyday of 'white' luggage, also known as 'rawhide' or 'vellum'. Reaching a peak of popularity in the early-to-mid 1930s, vellum luggage was mostly produced for use by women – hence the preponderance of surviving round hatboxes and vanity-cases. The ultimate baggage for the female Hollywood star was a full suite of matching, creamy-coloured vellum cases including a huge wardrobe-trunk with hanging space for dresses on one side and a chest-of-drawers for accessories on the other; at least one pair of large cases for outerwear; a blouse-case; a brace of hatboxes – one round, for large-brimmed straws, one square for a variety of less fragile hats; a shoe-case, and a rectangular vanity-case. These were wildly impractical, almost always bearing the evidence of the blue chalk cross used by railway staff to check off luggage, which left little or no trace on any other type of leather. Now scarce and often very dirty, at the time they

Opposite
Selection of red-morocco luggage. Superb vanity-case by Asprey with integral but separate pull-out jewel-case in the base (hence the 'fall-front'), on top of a Louis Vuitton red-morocco overnight-case. The small, unmarked jewel-case (*front*) is of standard quality.

Crocodile desk or deed box *c.* 1880; gharial head mounted as wall bracket; rare 'Hardy' fishing-fly wallet in crocodile hide *c.* 1900; crocodile-cased 'Voigtländer' camera (German) *c.* 1930.

exuded luxury, extravagance and frivolity, their glamorous image reinforced by the rapidly growing force of cinema, and for a few years they were as fashionable as the Filofax was in the 1980s. Occasionally one does find pristine vellum items that have been carefully preserved by a fastidious owner – when opened, often releasing the fleeting but unmistakable fragrance of lavender and talc, toilet soap of the finest quality and Eau de Cologne.

In 1934, a separate section was given to 'Chamois and White Leather Manufacturers' in the Post Office Trade Directory for London, with around seventeen companies advertising this speciality. It seems to have been produced from a variety of domestic skins. Bleached cowhide ('rawhide'), goat, sheep and pigskin were all used. If pigskin has been employed, the pores of the skin will be discernible, sometimes bearing an off-putting resemblance to human skin. However, when finely and sympathetically worked, unbleached and tanned, it can be attractive, and the material itself has more long-term durability than other types of leather, though its surface bruises easily.

Opposite
Canvas suitcase *(top)* and cabin trunk *(middle)*, both typically English, *c.* 1895–1935, on top of typical continental canvas trunk. Note how design of locks differs from those on the English examples. This particular item is probably German, since the applied metal tag reads *Garantiert echt papelplatten* – 'Guaranteed genuine fibre foundation'.

Selection of white-leather (vellum) luggage, typical of the 1930s, including two Revelation brand expanding cases *(bottom middle)*.

Over the years, a number of synthetic materials have been created for use in the production of durable luggage. As early as the 1890s, a prototype utilising compressed fibre of vegetable origin was employed to manufacture lightweight and comparatively cheap suitcases, hatboxes and trunks, and early this century a 'vulcanised' fibre was developed. There were many brands, the lesser examples of which may not even bear a maker's name, usually being stamped with the words 'vulcanised fibre'. The two outstanding ones, which have survived supremely well are 'Orient Make' and 'Pukka'. 'Orient Make', immensely popular by the 1920s and 1930s, boasted a subtly decorated lining of an extremely high standard. 'Pukka', formerly stocked by Harrods and the Army & Navy Co-operative Stores Ltd, amongst other exclusive retail stores was, along with the earlier canvas, hide-trimmed luggage, deemed far more suitable for tropical use than leather luggage, which stands up less well to extremes of heat and aridity. 'Pukka' luggage came with an individually numbered guarantee bond (see illustration) that promised: 'FREE REPAIR or FREE RENEWAL if beyond repair of the said article at any time within FIVE YEARS from date of purchase, subject to the conditions printed on back thereof'. The bond illustrated is for a trunk purchased on 30 October 1925; seventy years on it is still in excellent condition.

Aero alloy was another strong, lightweight material frequently used in cases from the late 1920s onwards, particularly for travel by air (see illustration on p.13). Wicker and raffia were also employed for

The 'Pukka' Luggage Guarantee Bond.

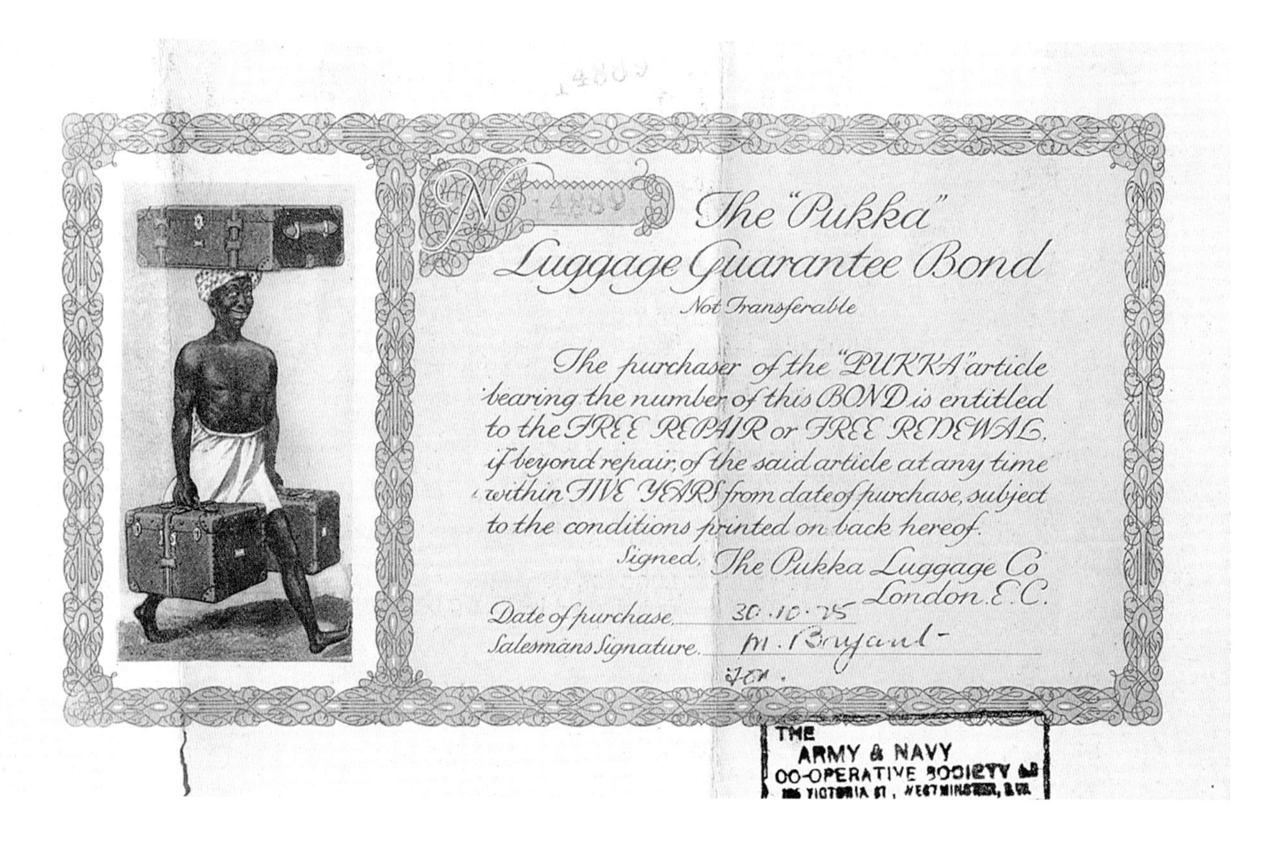
No. 14889

The "Pukka" Luggage Guarantee Bond

Not Transferable

The purchaser of the "PUKKA" article bearing the number of this BOND is entitled to the FREE REPAIR or FREE RENEWAL, if beyond repair, of the said article at any time within FIVE YEARS from date of purchase, subject to the conditions printed on back hereof.

Signed, The Pukka Luggage Co London E.C.

Date of purchase 30.10.25

Salesmans Signature M. Bryant

THE ARMY & NAVY CO-OPERATIVE SOCIETY Ltd VICTORIA St, WESTMINSTER, S.W.

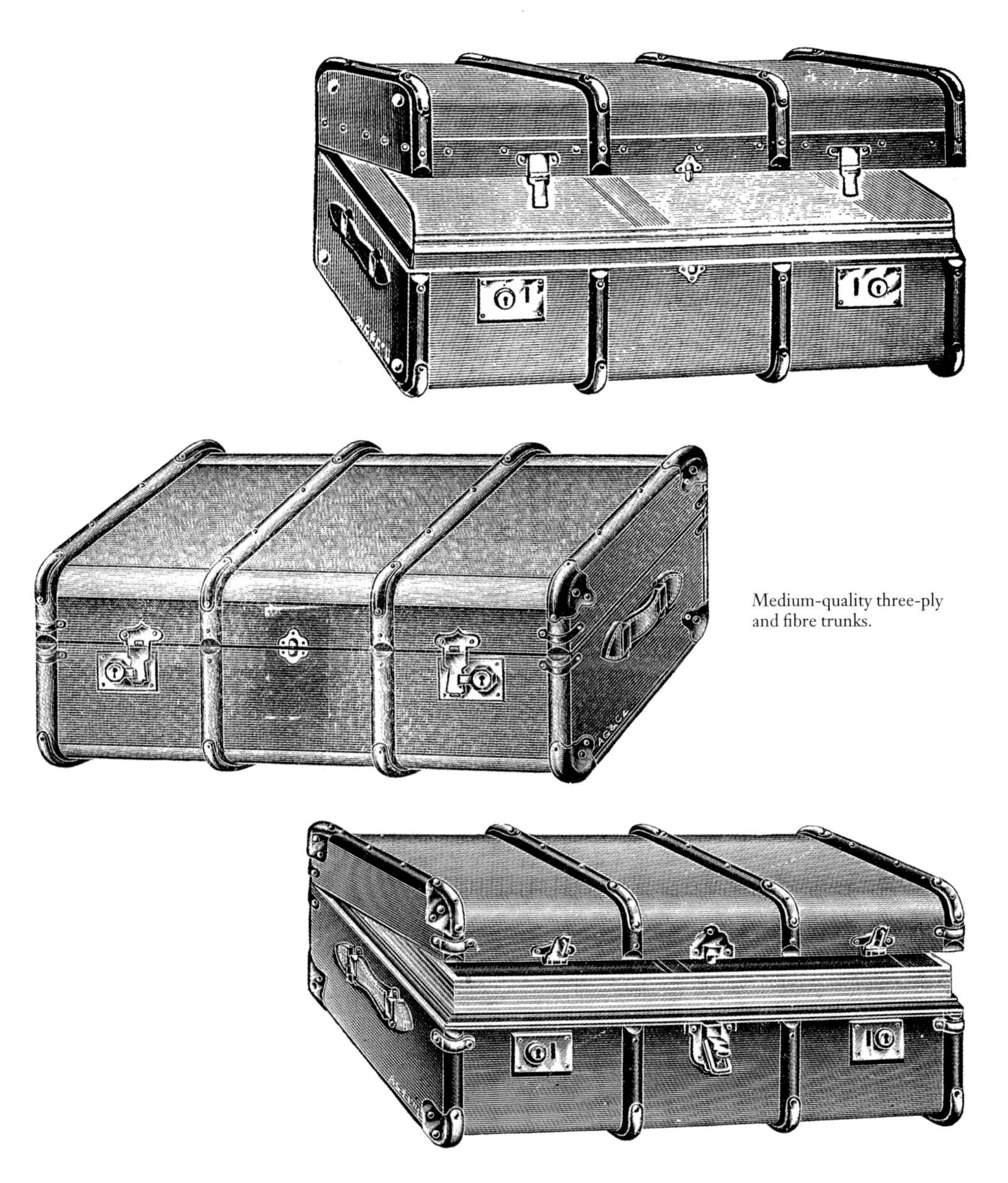

Medium-quality three-ply and fibre trunks.

this purpose but have rarely survived intact due to their tendency to dry out and become brittle with age. A few stunning examples have been preserved, and are highly sought after.

Equally rare are examples of pine and mahogany cases, generally of continental manufacture. Round, plywood hatboxes, which are currently being reproduced cheaply and to a high standard by a company called Greaves & Thomas (Brentford), were formerly made in Czechoslovakia and the Balkan States by small, family-run firms. The maker's name (if present) on an original will be rubber-stamped fairly large in purplish-blue ink on the interior of the lift-off lid. Approximately a third of the original examples had a decorative fretted white-metal edging applied around the lid. Absence of this element by no means rules out originality, but it is not present on the reproductions.

In cheaper luggage, a plethora of different types of leathercloth were utilised, and a multiplicity of trade-marked fabrics were employed including 'Hidex', 'Leatherette', 'Rexine', 'Dermide', 'Empire Leathercloth', 'Durexcloth', 'Rexhide', 'Excelsior', 'Levrine', 'Nacoid Brand' and 'Tannoid'. Such items, most widely used for luggage in the 1920s and 1930s, were rarely marked by the maker.

We have seen that many different hides and other, predominantly organic, materials were used prior to the 1940s in the manufacture of luggage, and that much variety and ingenuity was employed in the creation of these receptacles, both practical and beautiful. But, despite the breathtaking selection available, cowhide remains the favourite option.

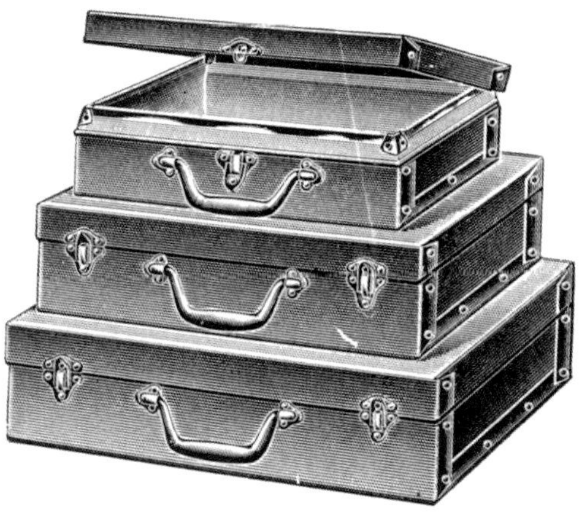

4

Lifting the Lid

The inside of a suitcase can tell the collector as much as its exterior. It may be lined with soft, gold-tooled, caramel, dark-green, deep-blue, scarlet, crimson or black morocco leather. This is usually an indication that it was made over ninety years ago. Plain morocco linings in reds, green, navy or royal blue and a vibrant shade of vivid purple came into vogue again in the mid 1920s, most often for writing-cases. If the lining is of restrained, dark-blue pinstripes on off-white or cream ticking, one may also be fairly certain that it is of Victorian manufacture. First-quality, old portmanteaux and Gladstone bags were lined with solid, gold-tooled morocco leather of smooth or oak-grain red, black, navy or green. Second-quality are still an absolute delight, lined with blue-on-cream-striped ticking, all edged around with a delicately stamped-out, finely crenellated red-morocco trim.

Pigskin lining, gold-blocked with impressed makers' details and gold-accented tooled decoration was another favourite option. Crocodile linings in fitted dressing-cases and writing-cases, the exterior of which were also crocodile in most instances, are a rare and sought-after possibility. A very few were lined with lizard or snakeskin; one of the oddest, produced by Cleghorn of Edinburgh, had a chunky, dark-blue, shagreen-stamped leather exterior and a natural, grey-and-white snake-scale interior. This is such a departure from the usual products made by the firm, which were of restrained, classic tan leather and fine pale-grey and green canvas, that it must have been specially commissioned by a purchaser wishing to possess a unique case.

Lady's finest crocodile travelling-case with sterling-silver fittings, from the 1934 Mappin & Webb catalogue.

Illustration of a medium-grade wardrobe-trunk, lined with a typically garish cretonne of the early 1930s.

Fabrics were also used, including Moiré silk, which came in many hues, occasionally with directly gold-blocked maker's mark. Today, similar quality silk would have to be specially ordered, and may cost upwards of £120 per metre – often more than the cost of the case itself. Most cases lined in dark but brilliant red and blue velvet will be small and distinctively shallow. Almost without exception, these will have started out as Masonic cases, and are often found complete with regalia and other small ephemeral items, such as programmes for Masonic dinners. For a relatively brief phase, mid-Victorian, bucket-shaped hatboxes for ladies' riding hats or gents' black silk top-hats were lined with cotton-velvet of red, blue or green. Due to the instability of many dyes of this era, they tend to have undergone an elegant process of fading. Hatboxes lined with black velvet can be dated to within a few years after 1861, the year that Prince Albert died, when many sentimental subjects,

Opposite
Jewel-cases and attaché-cases from the 1934 Mappin & Webb catalogue.

MAPPIN & WEBB

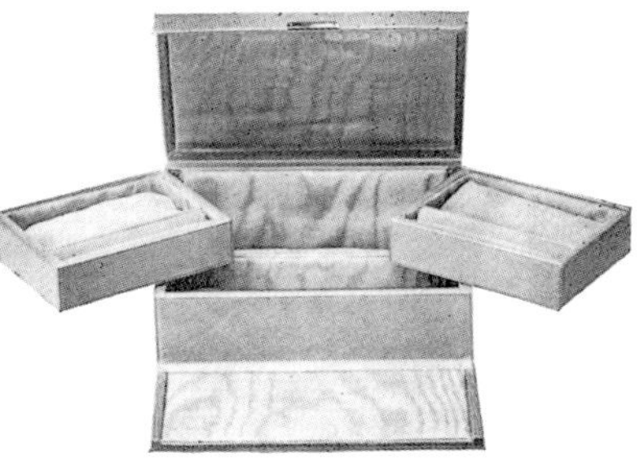

G **2245**.—Crushed Leather Jewel Case in smart bright colours. Lined Velvet and with deep compartment in Base for Beads, etc. Size when closed, 7 in. long × 5½ in. wide × 3 in. deep **£1 15 0**

G **2258**.—Bridge Case, lined Leather, with 2 Gilt Quadrant Hinges. Complete with 2 Packs of fine quality Playing Cards, Scoring Blocks and Rules.

Covered Crocodile	**£4 4 0**
Fine Seal	**2 17 6**
Fine Morocco	**2 10 0**

G **2267**.—Attaché Jewel Case, lined Velvet. Bramah action Gilt Lock. 10 in. long × 7 in. wide × 2¾ in. deep.

Fine Seal	**£6 15 0**
Fine Crocodile	**9 15 0**

G **2097**.—Flat Jewel Case, lined Velvet with lift-out tray and spring Bramah lock.

	Morocco.	Crocodile.
6 in. × 7 in.	**£3 15 0**	**£7 5 0**
8 in. × 9 in.	**4 5 0**	**8 5 0**

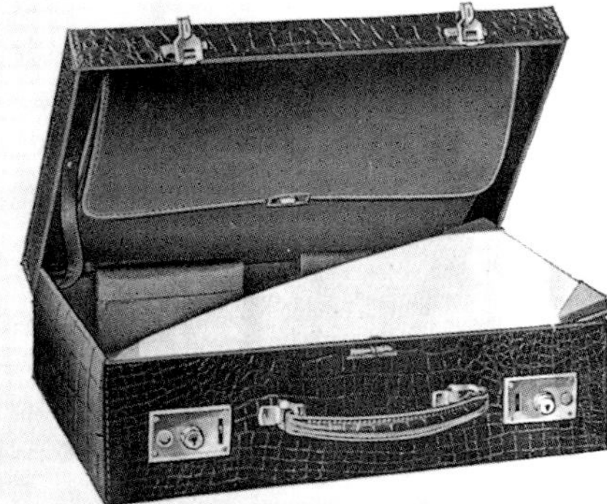

G **2018**.—Finest Crocodile fitted Attaché Case, lined crushed Roan. Size 15 in. × 10 in. × 4 in.
£12 10 0

G **2019**.—Lined Real Pigskin .. **£14 10 0**

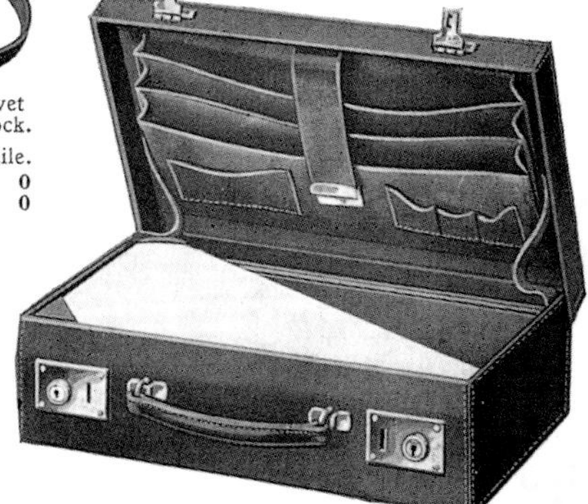

G **2259**.—Fitted Attaché Case in Hide, lined Leather. Size 15 in. × 10 in. × 4 in.
£2 15 0

LONDON SHOWROOMS:

2, QUEEN VICTORIA STREET, E.C. 4 — 172, REGENT STREET, W. 1 — 156-162, OXFORD STREET, W. 1

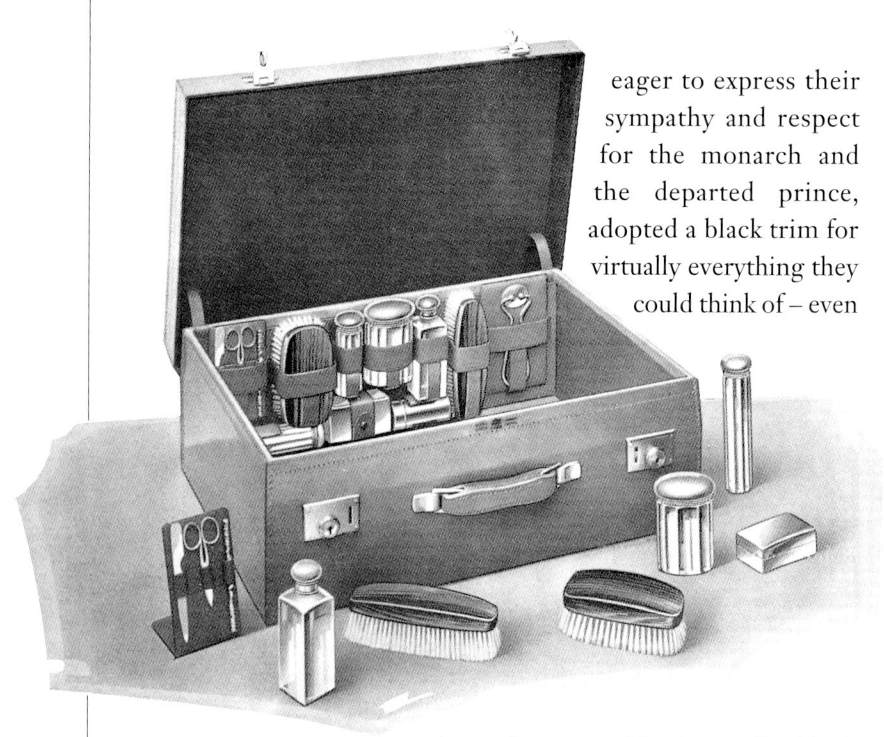

Gentleman's dressing-cases from Mappin & Webb.

eager to express their sympathy and respect for the monarch and the departed prince, adopted a black trim for virtually everything they could think of – even going so far as to add a decorative black band to the surface of tables.

Other items likely to have been lined with velvet are jewellery-boxes and cases; these are obviously made for the purpose and contain divisions and usually a section for rings. There will often be a compartment with a shaped, circular space, which can be a puzzling feature: this was for storage of one's pocket-watch, used by both men and women before the advent of the wristwatch.

Often found inside the mid-budget, vulcanised-fibre, 'Orient Make' cases of the 1920s and 1930s, discussed in the previous chapter, was an unusual lining of top-grade unbleached linen. The words 'Orient Make' and a distinctive rising sun logo, are jacquard-woven into the fabric, forming a motif about 3½ inches across and 6 inches deep in the right-hand lid section. The effect is charming.

Linen, and sometimes canvas in different grades, were fairly cheap, standard options for linings. The late 1920s and the 1930s saw many boldly extravagant floral prints and geometric designs, often extremely garish. Numerous vellum cases of the mid 1930s are lined with figured art-silk – usually a three-bladed propeller logo embossed on cream, *eau-de-nil*, burgundy or royal-blue silk, spiralling across the fabric in all directions, suggestive of the increasingly viable air travel.

Paper was also often used as a lining, most commonly printed in imitation of leather, but also in many different decorative patterns, which unashamedly announced themselves as printed paper. Embossed paper 'mock-croc' was a particularly popular choice in suitcase-style document-cases from the early 1930s on.

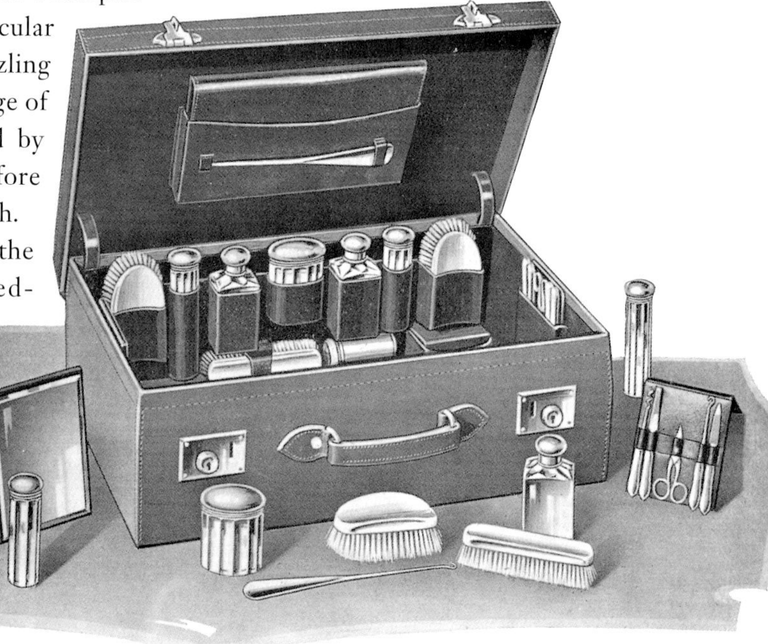

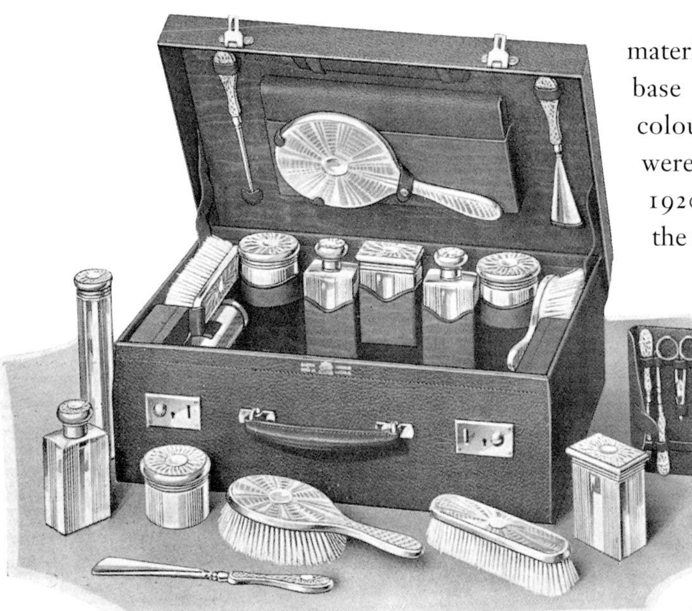

Lady's morocco dressing-cases from Mappin & Webb.

material, which dropped neatly into the base of the case. Tan leather and coloured morocco linings for these were greatly in demand from the mid 1920s to the mid 1930s. Sometimes the divisions are labelled for function in gold-blocked lettering: 'Answered', 'Unanswered', 'Bills Paid', 'Bills Unpaid'. Mock-crocodile leather lining was also occasionally employed.

It is fairly simple to ascertain from the fittings of a dressing-case whether it was intended for a man or a woman. A mirror with a tapered, integral handle was meant for women, designed to facilitate the process of checking the back view of the hairstyle against the dressing-table mirror. These were usually backed with silver, ivory, wood (ebony was a favourite) or, rarely, with tortoiseshell or shagreen.

Gentlemen's shaving mirrors, on the other hand, were normally of rectangular, bevelled glass, leather-framed and mounted, with a fold-out stand – rather like that on a photo frame – of

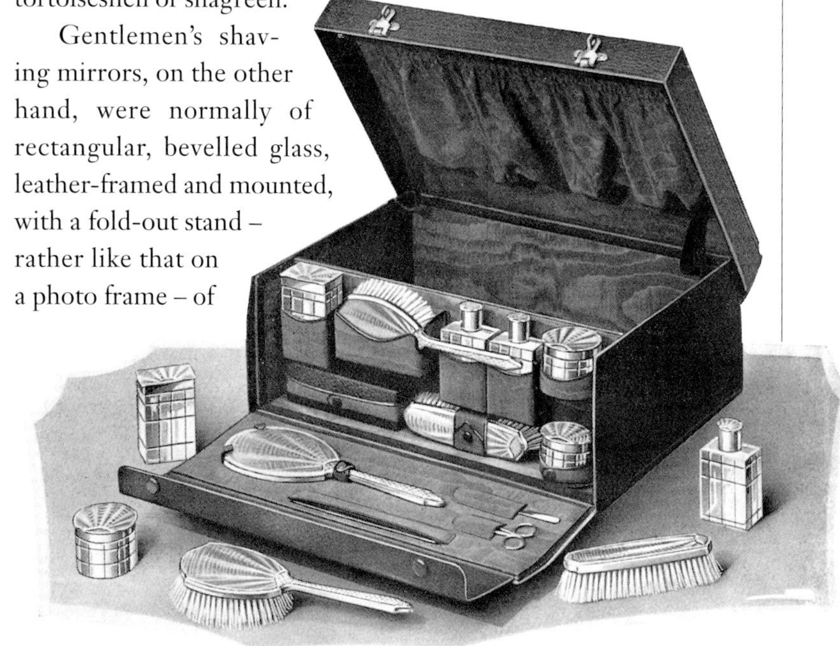

Another common lining was Rexine, or leathercloth – basically a strong, coated cotton or linen, generally in a shade of brown, but sometimes green, blue or red. 'Skiverine', 'Suedine' and many other types of imitation, or pressed and stamped leather substitutes were utilised as linings for cases.

Many fabric-lined cases had as an integral part of the lining little gathered pockets of the same material, designed as a flexible system to stow bottles or other small items snugly so that they would not rattle around. Generally, these pockets are found in ladies' blouse- and dress-cases rather than in the slightly larger-sized standard suitcases. Plain art silk was a favoured and feminine option, available in many hues from the palest cream, through rose pink, to various shades of blue and green.

Attaché- or writing-cases were often given as presents. They were frequently leather-lined, with divisions in the lid section for paper, envelopes, pens etc., and a separate blotter covered in the same

L 1969.—Lady's Black Morocco Dressing Case, size 18 in. × 13 in. × 6 in. Containing a Service of Sterling Silver and Fine Black Enamel Toilet Requisites of exclusive design, and Engraved Glass Toilet Bottles and Jars, mounted to match. Complete with Waterproof Canvas Cover.

£65 0 0

L 1980. — Lady's Blue Morocco Dressing Case, size 18½ in. × 13 in. × 6½ in. Containing a Service of Sterling Silver and fine Blue Enamel Toilet Requisites of exclusive design, and Engraved Glass Toilet Bottles and Jars, mounted to match. Complete with Waterproof Canvas Cover.

£67 10 0

LONDON SHOWROOMS:

172, REGENT STREET, W. 1 — 156-162, OXFORD STREET, W. 1 — 2, QUEEN VICTORIA STREET, E.C. 4

wood, metal or leather-covered card. Metal stands were often fashioned in a loop so that the mirror could alternatively be hung up. Again, ladies' hairbrushes would have a long, thin handle, whilst gentlemen traditionally used a military (oval or rectangular) pair. Up until the late 1920s, ladies generally required a buttonhook for their footwear. The presence of these provides yet another clue to the gender and era of the original owner of the case.

Most cases of quality, for either gender, also include a pair of clothes brushes, longer and narrower than hairbrushes, and occasionally a shoehorn. A small folder containing nail files, cuticle tools, scissors, tweezers, buttonhooks and bodkins was also generally provided. Additionally, in earlier cases, there is often a miniature corkscrew for opening the plain little corked bottles in which pharmacists sold cosmetics and scents before the advent of product branding, when manufacturers began to commission attractive individual containers specifically for their different products. Until this time, one would decant the contents into one's own ornate – ideally, vermeil-topped (gold-on-silver) – crystal bottles, ready and waiting in the dressing-case.

A particularly unusual example of a gentleman's dressing-bag is one bearing the maker's name of The Alexander Clark Co. Ltd (London). Masculine and imposing, it is stuffed full of solid-silver, English Assay Office marked bottles, each engraved with the name of the product it was intended to contain: 'Eau de Cologne',

Opposite
Mappin & Webb 1934 catalogue illustration of morocco dressing-cases.

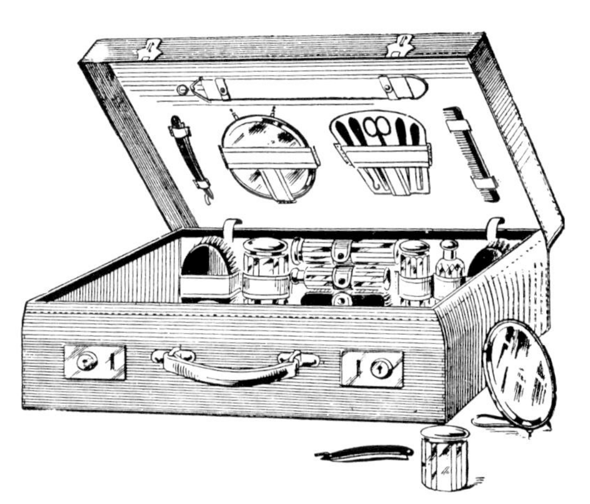

The 'Elite' – gentleman's fitted suitcase.

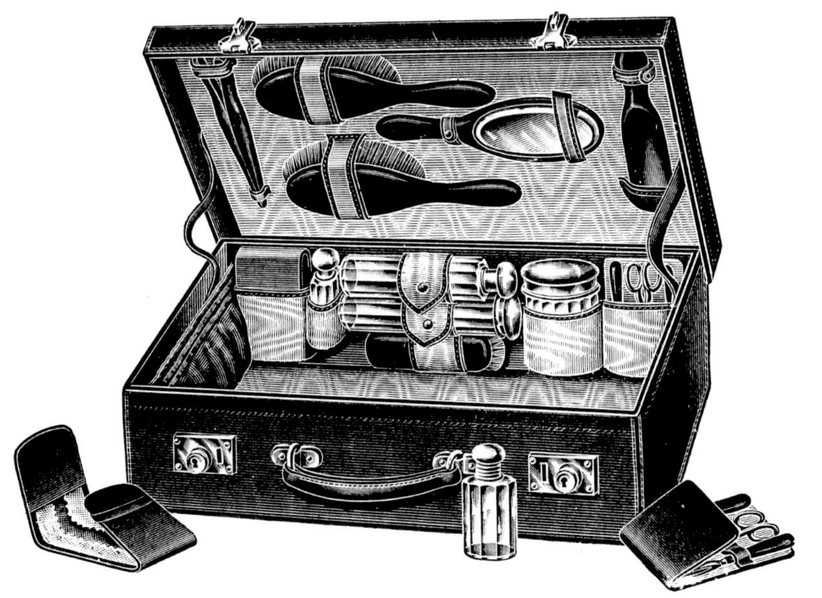

93
The 'Princess' – lady's fitted suitcase.

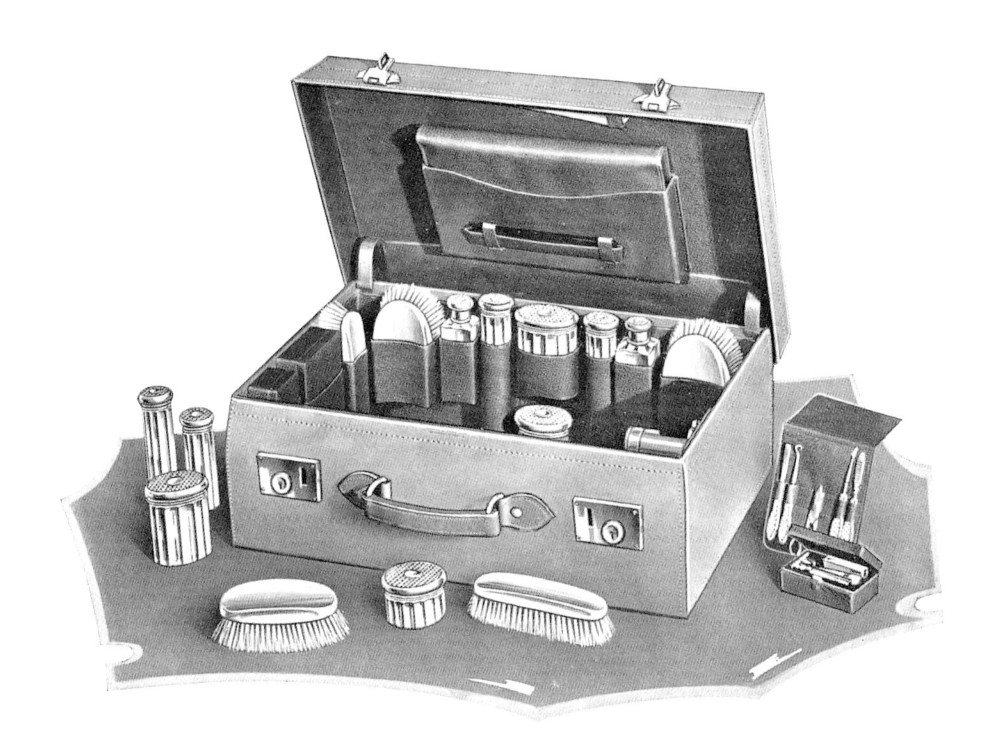

Gentlemen's dressing-cases in hide and crocodile, from the 1934 Mappin & Webb catalogue.

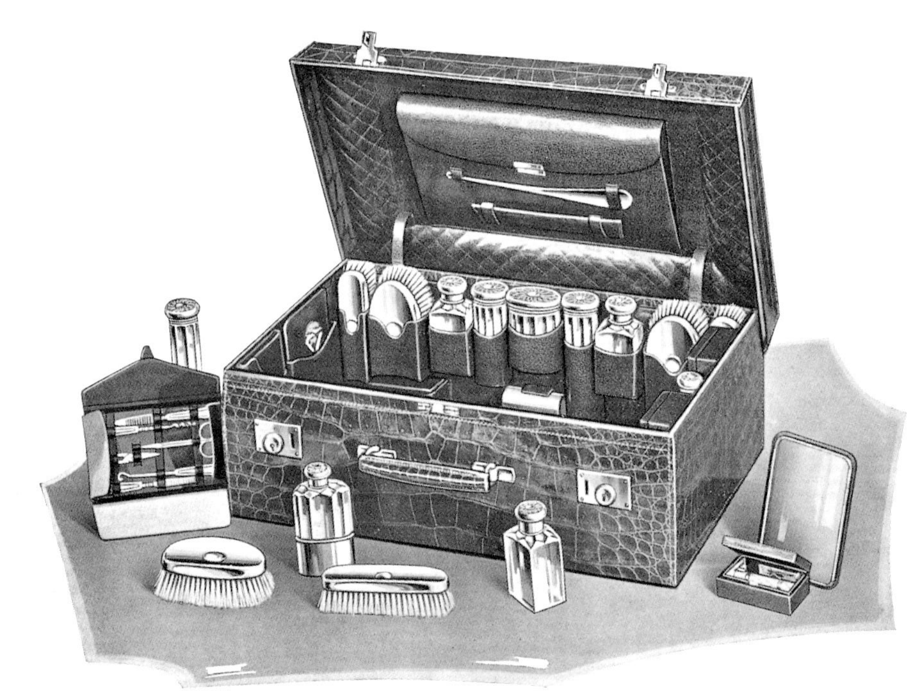

'Listerine' and so on. (Assaying is the process of analysing a precious metal. The presence of an official mark authenticates the standard of purity for items made of gold or silver.) This must have been a very special (and expensive) order: virtually all dressing bags and cases, even of the finest quality, contain silver-topped, lead-crystal bottles whose function is non-specific and thus flexible.

Little knick-knack boxes and tiny, travelling jewellery boxes; sewing sets complete with needles, pins and cut-threads in a folder; notebooks, diaries, miniature guides to etiquette, phrase books and dictionaries fashioned from the same leather as the lining of the case, may all be found.

Leather-covered boxes marked 'Light' or 'Vesta' for matches, complete with a flip-up metal striker in the lid or a metal striker on the base, came paired with a leather-covered bottle in a metal casing of exactly the same size and shape, often labelled 'Ink' – a travelling inkwell. The matches were intended for the purpose of either heating sealing wax, or for use in conjunction with another peculiar little gadget that may be found in a lady's dressing-case. This is a rectangular metal box, about 1½ x 3½ x 1¼ inches, which unfolds down the middle and reveals itself as a mysterious type of spirit stove. Its function is explained by the folding, metal curling tongs, often with ebony or ivory handles, which will probably accompany it – a precursor of the butane-powered travel tongs of the 1990s, vital to the 1890s coquette.

Glove-stretchers, resembling long-nosed pliers, sprung so that the tapered

Lady's finest crocodile travelling-case with solid-gold fittings, from the 1934 Mappin & Webb catalogue.

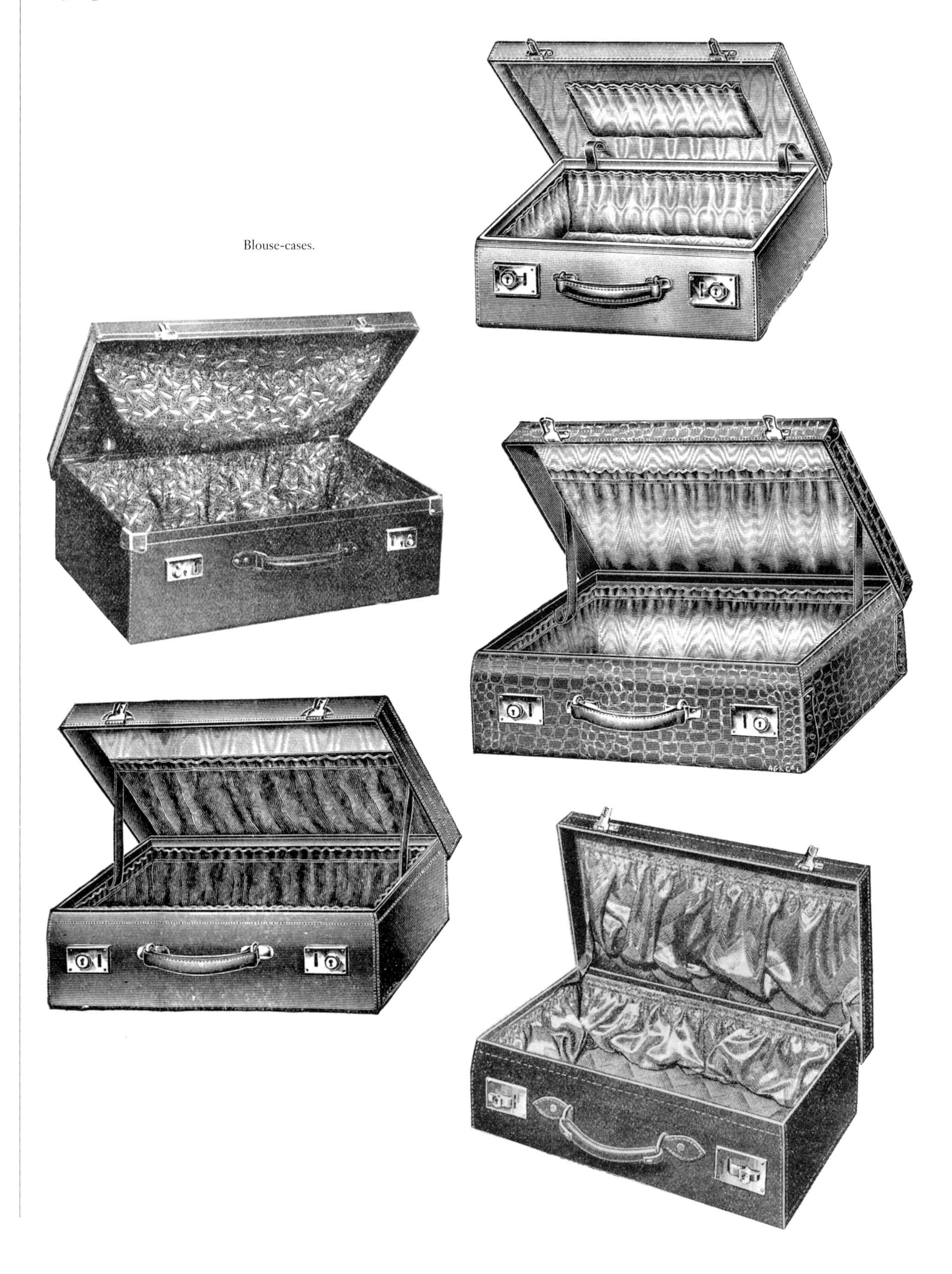

Blouse-cases.

tips could gently ease and expand the fingers of dainty new kid gloves when the flared handles were squeezed together, were a common inclusion in ladies' dressing cases until around 1910. These were usually made to match brush backs and other fittings, most commonly of ebony, sometimes of ivory or tortoiseshell, and were considered a necessity for the achievement of perfectly fitting gloves.

Prior to the mid 1920s, gentlemen were normally in possession of at least one cut-throat razor. After this time these items had evolved into compact safety razors of Bakelite or metal and were rarely included amongst the original fittings, since safety razors rapidly became as ubiquitous as the throwaway plastic versions of today.

Fittings were often made of silver or silver-gilt and, very occasionally, of solid gold. Shagreen, usually dyed green, but sometimes pink or blue in colour, was another luxurious possibility, frequently embellished with a monogram or initials of solid silver or gold. The manufacture of shagreen was revived by Ernest Betjemann, father of the poet John Betjeman (who dropped the final 'n' from his name) and popularised by Asprey & Co. Ltd (London) in the 1920s. Ebony and ivory, often, but not always, stamped 'Real Ebony' or 'Ivory' were fairly commonplace. Rarer, was Macassar wood, a type of ebony figured a little like rosewood, and latterly referred to as 'Bengal Wood'. Bottles and fittings without a stamped English silver Assay, or Hall mark are likely to be chromed, white metal or E.P.N.S. (electro-plated nickel silver) and this reduces their value dramatically.

It is unusual to come across a fitted dressing-case with its original contents intact. This is partly due to the great rise in silver prices in the late 1970s, when sadly, many people did, quite literally, scrap the family silver. Additionally, the individual items have an obvious intrinsic value and many possible functions. Dealers in antique silver will willingly purchase individual items from a fitted case, but show no interest whatsoever in the case complete with contents; the practice of 'breaking' a case has therefore been encouraged by market forces.

Finnigans suitcase with green-morocco lining, gold-blocked with maker's name *c.* 1925.

BUTTER

5

Maker's Marks

While there are many hallowed names in the field of antique luggage, it is by no means essential that a piece should bear a maker's name to be worthy of collection. Frequently, however, if an attractive item is thoroughly inspected, the trade stamp or label of a respected maker will be found. Failing that, the distinctive construction of a handle, or a particular style of lock fitting or lining may provide strong evidence, if not definite proof, of the original manufacturer. A few unsigned examples of products that are normally marked seem to have slipped through the net – perhaps at the request of the original purchaser, perhaps because an applied label was deliberately removed, or became accidentally detached.

Some items will bear both a maker's mark and a retailer's label. Both John Barker & Co. Ltd (London) and Harrods (London), formerly had their own trunk factories, in addition to supplying the luggage of other makers. As early as 1902, Harrods catalogues state that various trunks were London-made expressly for Harrods, and by 1915 canvas covers for cases and trunks were being produced by Harrods. The Trunk Factory appears to have become fully operational on completion of a factory building in Trevor Square, SW7, in 1921. By 1950 the Trunk and Saddlery Factory had moved to a building in Pavilion Road, London SW1, and production of luggage by Harrods continued into the 1970s. Items made in the factory will usually bear a gold-blocked stamp reading 'Harrods' in signature-style script, and 'London' or 'London S.W.' on the body of the case. The applied ivory or ivorene (imitation-ivory) label bearing the name 'Harrods' may be found on all types and many brands of luggage, from 'Pukka' vulcanised-fibre trunks, to picnic sets by

Opposite
Typical, black leathercloth 'Coracle' brand picnic set. This example was preserved in a canvas foul-weather cover and is in pristine condition.

'The Motor Visite – a strong convenient case, canvas covered.'

G.W. Scott & Sons (branded 'Coracle') and other makers, to 'Revelation' brand expanding suitcases.

The list of large department stores that supplied luggage in London alone is extensive, and many are still trading today, offering a wide variety of goods from a number of manufacturers. Selfridges, Army & Navy, Peter Robinson, Debenham & Freebody and Whiteleys are familiar household names that may frequently be found applied to fine examples of vintage luggage.

The retail labels of Selfridges and William Whiteley are mostly narrow (about 2 1/2 x 3/8 inches), imitation-ivory tags, stamped with the details in black or brown and attached with a tiny brass pin at each end. The trunk department at Peter Robinson appears to have favoured the ubiquitous oval, gold-stamped, red-morocco retail label, while the Army & Navy Co-operative Stores Ltd used a shield-shaped, gold-stamped, morocco label in a variety of different colours. On hide items there is almost always a small, impressed, oval stamp on the body of the case itself – usually to the front of the base section above the handle. An extremely varied range of luggage, from bucket-shaped leather and square canvas hatboxes, through canvas, leather and crocodile cases of every size, to a wide selection of trunks, may be found bearing this morocco escutcheon, approximately 1 inch square

Blue morocco-cased tea-set by Leuchars *c.* 1880 on 1920s shoe-case in dark-blue leathercloth. The latter has a strong, metal-mounted handle and loop for attaching the opened case to a hook on the back of the cabin door aboard ship, providing easy access to eight pairs of shoes.

Opposite
Dome-topped crocodile picnic set by A. Barrett & Sons, with applied silver crested 'G' on lid (unseen) and marked on each item.

A. BARRETT & SONS.
MANUFACTURERS
To the Royal Family
63 & 64, PICCADILLY.
LONDON.
SALT

Selection of green-morocco luggage. The bolster-bag (*back left*) is marked 'Asprey', and the large, silk-lined suit-case is marked 'Harrods London SW' – both gold-blocked. The small sewing-case by John Pound is, unusually, gold-stamped with the maker's name on the bottom exterior of the case.

and often red or green in colour (although black or brown are also found) with the stamp 'A&NCS Ltd'. Look out for distinctive cases and square hatboxes – some of cowhide, some of brown or pale-green canvas, leather-bound on lateral edges only.

Asprey & Co. Ltd have usually marked their products with the words 'ASPREY LONDON', frequently beneath a crown, gold-blocked onto the body of the case or lining. Unusually for luggage, the mark may be found on the back-interior of the base, or placed towards the lower part of the lining of the interior lid section. Sometimes the mark may simply read 'ASPREY', applied in later examples to the lining, generally stamped onto a small square piece of hide identical to that employed on the exterior of the article, and often placed, unusually again, in an asymmetric position towards the upper left-hand side of the lid section. Heavy-weight, oak-lined, honey-coloured hide picnic boxes (usually for four people or more) and motoring cases of the late 1920s to the late 1930s, frequently stamped 'ASPREY' in black, are much sought after. There are also examples of these distinctive pieces with only 'PATENT No APPD 29305' stamped in black, but to the attuned eye, these are typical products of Asprey & Co. Fittings, where still present, will usually be stamped either 'Asprey' or 'A & Co'. Such substantial items are likely to have cast-brass, nickel-plated, side carrying handles with a distinctive construction.

Asprey & Co., who have created many unique items and continue to do so to this day, have also produced magnificent, black leathercloth-covered motoring picnic sets of obviously superior quality, similar in appearance to those branded 'Coracle'. Nickel-on-brass fittings are usual on these heavy-duty items, designed for robust use outdoors, whereas items such as jewellery-cases, where it could reasonably be expected that great care would be taken to keep both case and contents protected from the elements, are most likely to have gilded-brass fittings. Silver corners and catches are normal on smaller, highly decorative items. The attention to detail in every respect makes this name highly desirable. Many of these products, especially of coloured morocco leather – vanity and jewellery-cases, dainty silver-mounted wallets, purses and so on – are made to pre-war standards even when of later manufacture, and survive in wonderful, even perfect, condition, lovingly kept in the vivid, purple cardboard box in which they were packed when sold. These are avidly collected.

A similar quality is apparent in the products of Gustave Keller and of W. Leuchars – indeed, in 1888, Asprey & Co. absorbed the London branch of the latter. The Paris branch of W. Leuchars was taken over by its former manager, Monsieur Geoffroy, who acted as the Paris agent for Asprey until 1902. Asprey also bought out several other companies known for leather goods, including Houghton & Gunn (London) in 1906, and the lease of the property of S. Last (London) in 1902. Edwards of King Street, London, was absorbed much earlier, in 1859, and Thomas Jeyes Edwards remained in the employment of Asprey & Co. until his retirement in 1872. Founded in 1817, Edwards of King Street were leading manufacturers of patented writing and dressing-cases, appointed Dressing-case Makers to William IV in 1832. The Company was awarded a Gold Medal in the Great Exhibition of 1851 for

an innovative gentleman's dressing-case, the first to be made of wood covered with leather. Before this development, writing and dressing-cases were constructed of wood alone. Edwards was thereafter awarded the Royal Warrant as dressing-case, travelling-bag and writing-case maker to Queen Victoria, and Asprey & Co. were awarded the Royal Warrant for the first time in 1862.

Edwards & Sons Ltd (of Regent Street, London, as distinct from Edwards of King Street), were Royal Warrant Holders in the early years of the 20th century as stationers, silversmiths, dressing-bag manufacturers, and by 1915, antique dealers. Companies that also worked in these related fields, including Asprey & Co., W. Leuchars and Gustave Keller (Paris), invariably seemed to achieve a superior fineness of workmanship and detail. Rare and exquisite dressing-cases and bags, often of crocodile, with the gold-blocked maker's name of Edwards & Sons Ltd bear out this statement. It is also true of such delights as the crocodile picnic set (see illustration on p.63) with silver fittings, gold-blocked in the interior 'A. Barrett & Sons', a London-based company that held the Royal Warrant as 'Brush and Leather Goods Manufacturers, Silversmiths etc.' Further names to be found gold-stamped on exquisite goods with this especially fine feel are W. Thornhill & Co. (London), J.C. Vickery (London), The Goldsmith & Silversmiths Company (London) and Mappin & Webb (London). The dressing-cases of the latter two, both holders of the Royal Warrant, are frequently found in superb condition, still snugly contained within their leather-trimmed protective covers.

Individual fittings, where original, are usually stamped with the corresponding maker's name. A standard reference book on silver marks will greatly assist the collector of dressing-bags and cases in establishing the authenticity of the fittings, and since the U.K. system of marking precious metals incorporates the use of a date stamp, it is frequently possible to identify the exact year of production.

Alexander Clark Co. Ltd, which was bought out by Mappin & Webb in the mid 1960s, was another fine maker of dressing-bags and cases, as well as of more standard suitcases. Solid-silver fittings will often enable the article to be accurately dated. Cases are usually stamped with the name in black, directly onto the body of the hide, either with an oval stamp similar to that employed by A. & N.C.S. Ltd, or else in a distinctive script.

John Pound & Co. Ltd is a highly collectable and prolific name, though slightly less well known than those previously mentioned. As well as making silver-fitted dressing-cases, they advertised variously as trunk and portmanteau makers (including trunks of compressed cane and fibre), and produced motoring touring trunks and a wide range of cases and bags. The name may be black- or gold-stamped onto the body of the base, directly impressed on each side of the lid on trunks, or on an applied label, sometimes of stamped metal. Locks are frequently stamped 'JP'.

Finnigans was yet another maker to produce a wide range of luggage of all types, and is equally collectable and revered. Along with W. Barrett & Sons and Lansdowne Luggage, they almost invariably stamped each lock with their name, and often 'London', 'Liverpool', 'Manchester'

Opposite
Cube-shaped camera-box on a typical selection of early 20th-century suitcases.

or 'London & Manchester'. The name and address(es) of the company are, again, almost invariably impressed on each lateral side of the lid section of suitcases and trunks made of cowhide. (Insall, Bracher's, and Ellenger & Co. of Newcastle on Tyne were other makers who routinely marked suitcases thus.) The attention to detail extended yet further to gold-blocking the name of Finnigans onto the interior lining, when of silk or morocco leather. It is not unusual to find an object, particularly a suitcase, by Finnigans bearing all three methods of marking. The huge variety of different styles of luggage made by Finnigans means that in one or two instances, the usual means of marking may not have been employed – a vellum suitcase, for example, may bear a gold-blocked stamp on its upper base section. Many Finnigans leather suitcases have very delicate handles in proportion to the size and weight of the case. This may have been an early attempt to apply the concept of 'built-in obsolescence' and must certainly have kept the company busy with repair work. Sadly, these unique and attractive cases can no longer be taken back for repair by the original maker and an increasing number are now either completely bereft of handle or repaired with the wrong one.

The name of Finnigans may be found on other styles of luggage, from the gold-blocked wallet in crocodile, through hatboxes and silver-fitted crocodile dressing-cases, to enormous, vellum wardrobe trunks. It is also found on their own brand of expanding luggage, and on 'Revelation' expanding suitcases – perhaps the sturdiest and most versatile to have been developed. When the case is empty the base sits snugly inside the lid, but when full, its sliding metal fittings – the only link between the two separately constructed sections – expand vertically to allow them to be secured increasingly further apart to meet the space requirements of the user – almost doubling the carrying capacity. A number of other manufacturers' labels including W. Wood & Son Ltd (London), Harrods, Lawrence & Sons (London) and H. Boswell & Co. (Oxford) have frequently been found attached to 'Revelation' suitcases. Production of this wonderfully durable and popular type of suitcase sadly ceased just over twenty years ago, and the brand name was sold. The highly practical design, though usually made in cowhide, was executed in every imaginable material, from cardboard and leathercloth, vellum, vulcanised fibre, to crocodile and, on at least one occasion, elephant hide. The metal fittings may be of nickel-on-brass, naked brass, or more rarely of white metal, and are frequently stamped with the trademark logo – a striding porter with limbs that echo the expanding fittings, carrying a suitcase. The label, applied to the lining of the case, came in a variety of designs: a diamond-shaped lozenge, or an oval, paper label of black or dark blue, giving patent details and the name, or sometimes a circular label, silvery-grey, with the Revelation logo cut out within the circle. A plain, square label in these colours is yet another type. The exterior base section of the case frequently has a small, imitation-ivory label attached to the front, near the handle, stamped with the logo in black. Other examples bear a long, narrow, applied, brass label stamped 'Revelation' in one of a number of styles of lettering. This type of case was produced in the early 1930s through to the mid 1970s.

Opposite, top to bottom
Attractive, much-travelled, English attaché-case, unmarked; case by J. Nigst & Sohn, Vienna; weekend-case with tooled decoration, no maker's marks; 'Standex' brand suitcase with integral luggage tag in the handle fixing ('Pat. applied for' refers to this feature); case by Paul Romand, Paris, formerly owned by Mrs J. Gordon-Duff.

E.H.
M.
CABIN
FYFFES LINE
B
L.G.
LUGGAGE

R
SHAW SAVILL
LINE

Similar in concept to the 'Revelation' case was the 'Crescent' brand, though far less plentiful and varied. The extending fittings seem always to be of naked brass, and are more elaborate, having decorative ridges and bumps. The body appears to have been made invariably of reddish-brown cowhide and, unlike the 'Revelation' brand (which was produced in a wide range of sizes and proportions) has been confined to a couple of standard suitcase sizes. A stamped-brass label on the front section, near the mid-point of the handle, reading 'Crescent', indicates this maker. Wholesalers such as A. Garstin & Co. Ltd (London) produced similarly designed cases.

Another type of expanding suitcase, in which the sides of the lid section are not completely structured and incorporate flexible bands of leather, which enable the capacity of the case to be increased, tends to have endured the ravages of time and use less successfully. Unless such a case is in excellent condition and of exceptional quality with a good maker's name and, moreover, of a desirable hue, it is unlikely to be deemed collectable. Frequently produced in lightweight materials, later examples of 'Revelation' brand expanding luggage may be of canvas or textured-fabric-covered fibre, fitted with light alloy locks. Their metal-reinforced handle renders them extremely durable and practical, if not quite as widely admired as the heavier, classic hide version.

Other lightweight brands created in the 1930s through to the 1950s are still in use today. 'Orient Make' vulcanised-fibre cases, with leather cap corners and a metal-reinforced hide handle, arched and stylishly rounded for user-friendliness, are

Opposite, top left to bottom
Lady's round hatbox *c.* 1910; gentleman's double 'bucket-shaped' hatbox *c.* 1870; black leathercloth top-hat box *c.* 1880 by Moritz Tiller & Co., Vienna, Austria.
top right to bottom
'Bucket' hatbox for a single top-hat *c.* 1870; lady's round hatbox *c.* 1920; black leathercloth, tan leather-trimmed, round hatbox *c.* 1915; white-metal lock stamped 'Eagle Lock Co – made in USA – Terryville, Conn.'

Rare Louis Vuitton luggage:
Top
'Sac Chauffeur'.
Middle
Pair of matched, hide motoring cases.
Left
Vulcanised-fibre attaché-case.

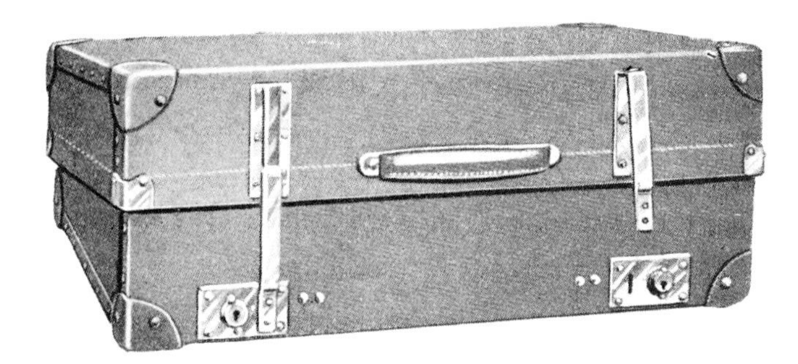

Expanding suitcases produced by Garstin & Co. Ltd.

generally practical for frequent use, if all too rare. An example in excellent condition may still bear the two pale-blue, black and gold transfer labels denoting the make, one to each exterior lateral side of the lid section, along with the rising-sun logo and brand name woven into the lining on the right-hand interior of the lid. The metal retail label of another company may also be attached, often near the rim of the base section at the back. Even the cardboard 'Travair' and 'Spartan' brands of the 1950s, if they have been well cared for, may still be fit for practical use. 'Travair' brand, although by no means of top quality and sometimes by now a little frayed in appearance (like a number of products in the lower-budget bracket), are not as stylish, often having a moulded-plastic handle. 'Python' was another such brand, generally of low-grade, brown leathercloth on a fibre base. None of these are what could be termed status symbols, but a surprising number are in frequent use by the modern traveller. 'Travair' brand cases have an oval, greenish-blue and silvered paper label to the interior lining of the lid section, centrally placed, whilst 'Python' brand are stamped black or silver to the exterior front rim of the base section above the handle, usually with the logo of a coiled snake. 'Spartan' cases may retain a red-and-black-printed, creamy-coloured, imitation-ivory label to the front exterior of the base section just below the centre of the handle.

The lightweight, metal-alloy cases produced for air travel towards the middle of the 20th century have often been stripped of their original painted finish and polished down to the shining metal beneath. This removes any trace of the applied-transfer maker's label. The earlier 'Astral' brand alloy cases (see illustration on p.13) made by the Heston Aircraft Co. Ltd (Slough) often have 'BRITISH PATENT NOs 165333/173997 MADE' stamped on each of the tin side-catches, and are of a distinctive, rounded shape, having a slightly bowed base and lid, with flat front and back. Originals were usually painted bottle green with two bands of leaf green around the centre, meeting at the middle of the handle. In later years, 'Astral' luggage was updated and sold in matching sets in '9 glorious colours': midnight black; Oxford or Cambridge blue; grenadier red; maroon; autumn brown; deep ivory, and nile or laurel green. This development (advertised as 'light as air to carry – streamlined in design') is even less frequently found with its original painted finish. A similar brand, 'Airport' luggage (see illustration on p.13), was normally produced with a painted surface in one of several shades of brown to suggest leather, with a round transfer label (and matching larger sales tag at point of sale) in red, blue and white. Again, these are often bereft of their original finish, and it is therefore a thrill for the purist to come across an untouched article not tampered with by dealers with the magpie instinct for gloss and shine at the expense of originality.

Cases manufactured specially for use by the motorist are both rare and collectable, most particularly to those fortunate enough to be in possession of the right type of classic car to put them in. J.B. Brooks & Co. of Birmingham produced motoring picnic sets, generally covered in black leathercloth, durable and virtually waterproof. Their motoring cases were frequently supplied in sets within a larger, bulky retaining trunk, into which they slid like drawers. These are often incomplete, as the cases were practical and fairly lightweight and extremely durable – often, no doubt, retained for use individually by an owner no longer requiring the remainder of the set, and unwilling or unable to store the bulky retaining trunk. Therefore, it is also fairly common to find individual cases, and with much patience it may be possible to re-marry a separated set. This is by no means an easy task, however, and the quest is only worthwhile if the first-discovered component is obviously exactly what is needed to complement a specific vehicle.

Zinc-covered, brass-bound, Louis Vuitton trunk for tropical use.

S. Reid Ltd (London) made distinctive sets of motoring cases with sloped fronts in hide and also in leathercloth. The latter were usually black with leather trim, stamped with the maker's name on a shaped handle with rounded tapered ends, which lay flat when the case was not being carried, held under two cast-metal keeps. These cases often have handles on each side, enabling them to be easily carried by two people. Found on examples of motoring luggage from the first third of the 20th century (and even later on cases by Karl Baisch, Stuttgart), side carrying handles were borrowed from a traditional design for large, heavy leather trunks. Shaped motoring cases with their charming sloped fronts are rare and collectable, even to those without a suitable vehicle. Pairs of these cases, one slightly smaller than the other, are particularly interesting, and an even rarer trio of such cases, graduated in size, is guaranteed to create interest, especially if they are made of cowhide.

The very different cases made specifically for Mercedes-Benz cars by Karl Baisch of Mercedeßtrasse, Stuttgart, again usually in sets of three, have similar handles of pale-tan leather, and are bound around the lateral edges with leather in a matching shade. The covering of leathercloth may be black or brown, and the lining is usually a checked or tartan-type design in toning shades of brown, white and creamy yellow. These are rare and highly prized, especially when found in complete sets, though again, they are most often discovered individually. Shaped sets of cases for V.W. beetle and Citröen cars are equally legendary and just as difficult to locate.

The large cases and Gladstone bags produced by companies such as H. Greaves of Birmingham (see illustration on p.107), made on a metal frame and lined with morocco leather, with the maker's name stamped in gold onto the lining, may be too weighty and, by now, a little fragile for the rigours of modern travel. Nevertheless, their decorative appeal is increasingly obvious as time goes by and fewer objects of such outstanding traditional workmanship are produced. Another maker whose bulky, old-fashioned products have recently been reappraised are H.J. Cave & Sons (London). The name is now highly prized, particularly on the type of heavy cowhide shirt-trunk with interior, linen-covered drawers that is now recognised as a stylish and attractive design. Allen's Portmanteaus (London) produced similarly cumbersome trunks and bags, and are now once more highly appreciated for their superb craftsmanship and the quality of the hide used in their manufacture.

Two fairly prolific West Country makers, exceptional both for the quality of hide they employed in manufacture and for the distinctive style, strength and suitability of the handles they created, were Insall & Sons and Brachers. The cases were usually of top-grade, heavy cowhide, although it is possible to find an occasional early leathercloth example – of academic interest, but by no means as commercial today as the gloriously patinated leather version. These two companies were apparently at their zenith in the late Victorian/early Edwardian era, and their products are strikingly similar in appearance. In addition, their suitcases share a relatively rare design flaw, being produced mainly without 'keeps' – either a strip of fabric or leather fixed to the mid-side of the lid at one end and to the mid-side of the base at the other, or more commonly, two strips of leather each riveted to the upper back base at one extreme and the lower lid at the other. Nearly all cases by other makers have this feature, which prevents the user from inadvertently bending the lid section too far back, leading to the lid and the body of the case parting company where the leather has split. Consequently, it is very rare indeed to come across a case by either Insall or Brachers in entirely original condition: they are likely to have an old 'strip' repair to the back seam. This will, by now, usually have mellowed to an acceptable match with the original hide, and even the repair is likely to be older than most luggage one comes across.

These cases are among the loveliest, still functional, but genuinely antique (by the time-honoured definition of 100 years or older) that one is likely to discover. Some are lined in caramel-coloured hide, unbleached linen, red morocco, or even mock-crocodile. Expect to find a leather-edged flap in the lid, and accompanying leather straps, along with another pair of retaining straps in the base. There should be a gold-blocked, leather label of tan hide or red morocco, usually oval but sometimes rectangular with canted or rounded corners, placed centrally onto the interior lid flap. Even if this is missing, the side sections of the exterior of the lid will invariably bear the maker's mark centrally, impressed directly onto the hide in an oval cartouche. Another, rarer and earlier maker, whose products can now be called antique by any definition, was Webb & Bryant of Portsea. Hatboxes and cases stamped with this name are notable for the beautifully patinated cowhide from which they are formed.

Top to bottom
Cases by Insall; Finnigans; Brachers (note similarities with Insall case); John Pound.

Insall & Sons were Bristol-based, while Brachers labels state 'Bristol & Cardiff'. Other Welsh makers and suppliers are rare, though H. Moores, Saddler, of Cardiff is a name occasionally found. Cases marked by Irish makers also appear all too rarely, at least on the London market, from which perspective this account is unavoidably written. However, many examples of luggage by prominent Scottish makers such as Cleghorn, A. Boswell, Forsyth and Irving Brothers, all of Edinburgh, may be discovered, along with the products of Leckie Graham and Reid & Todd, both thriving for many years in Glasgow. The traditional appreciation of quality in Scotland appears to have ensured a prolific output of exquisite leather goods by the premier Scottish makers. Cases by Cleghorn are often of cowhide, but the leather-trimmed canvas luggage bearing this maker's name is of exceptional quality and charm. Most of the Scottish makers appear to have stamped their brand name in black onto the body of the case. Many cases by Forsyth are of cowhide, the rim of the base uniquely trimmed with crocodile. Reid & Todd of Glasgow advertised their luggage and umbrellas with equal pride.

Another company to have produced umbrellas and leather goods, along with whips and sticks, was Swaine & Adeney (London), later incorporating Brigg, a famous umbrella maker, to become Swaine Adeney Brigg. Mainly known amongst antique dealers for fine riding crops, to the luggage collector the company is remarkable for those lovely little cases, approximately 6 x 7 x 1½ inches that could be attached to the saddle of one's horse, and which transported a snugly fitting flask and sandwich tin. The finer examples contain a silver-topped crystal flask and solid silver tin; those of lesser quality an E.P.N.S. flask and tin. This basic design has been executed by other makers, but never better than by Swaine Adeney Brigg. The company still produces luggage and fine leather goods.

Drew & Sons (London) are primarily renowned for their wicker 'En Route' tea baskets, fitted with custom-produced and distinctive kettle, burner, stand, tins and crockery, each and every piece proudly emblazoned with the maker's name, down to spirit container and cutlery. Detailed instructions for use are printed chocolate-brown on the buff leathercloth stitched into the lid. The fittings were most often of silver-plated metal, but in the very rare, more expensive examples, they are of solid, English silver, typically dated *c.*1908. Despite a little inevitable but restorable distress to the aged wickerwork basket, given their obvious utility, beauty and romantic appeal, these vintage wicker items may still be regarded as desirable. Many leather-cased and black leathercloth picnic sets by Drew & Sons also exist, as do a wide variety of superb crocodile, silver-fitted dressing cases, trunks, hatboxes and attaché cases. Non-wicker picnic sets usually have an ivory label stamped 'Drew & Sons'; leather-cased sets may bear an impressed stamp on the body of the case in addition to, or instead of, the applied label. Because of the variety of items manufactured, luggage may be marked either with a gold- or black-blocked mark impressed directly onto the hide, or sometimes with an oval, morocco label. Together with the luggage produced by Finnigans and John Pound & Co., those items are well-known and much desired by collectors.

An extremely famous brand name, which can have escaped the notice of few, is Louis Vuitton. A potent status symbol, its yellow monogram on brown canvas is instantly recognisable throughout the world. However, what may be less widely realised is the fact that the company has produced an astonishing variety of luggage in almost every material in its long history. Striped, and later, chequerboard-patterned canvas (named 'Damier' by the company, and recently re-launched) became so widely imitated that the monogram became necessary to establish authenticity. Earlier trunks and cases produced by Louis Vuitton are covered in a canvas into which the

Top to bottom
Rare, very early Louis Vuitton hatbox in striped-fabric, with pre-patent lock; English milliner's painted-wood hatbox, late Victorian (note mimicry of above design); large Louis Vuitton hatbox in 'Damier' canvas, colloquially known as 'chequerboard' fabric; very rare Louis Vuitton fabric trunk with pre-patent locks (the red-striped toile was first produced in 1872).

DREW & SONS
DREWS
DREWS PATENT.
DIRECTIONS
DREWS'
"EN ROUTE"
TEA BASKETS.

pattern was woven and, though the nature of the material has changed over the years, to keep abreast of technological developments, the unique character of the monogram design has been retained. In addition, down the years, items of luggage in many plain colours have been produced to order. Orange, primrose-yellow and plain brown canvas 'Vuittonite' are easier to find than the more unusual colours, some of which were reserved for specific customers. Smooth hide, crocodile, morocco leather and even vulcanised fibre have also been used in the manufacture of Louis Vuitton luggage. Since 1854, whilst offering an extensive choice of ready-made items, the company has designed and made special orders. Amongst these was a trunk with special louvered trays to carry fresh fruit, designed for Ismail Pasha, Sultan of Egypt; a tea-case for the Maharajah of Baroda; a 'secretaire' trunk for the conductor Leopold Stokowski, and the 'Milano' case, an ornate item lined with red-morocco leather, which was displayed at the Paris exhibition of Decorative Arts in 1925.

Louis Vuitton's picnic sets, and dressing-cases with many silver-topped bottles decorated with a simple band of continuously repeated 'LV' initials around the lid, are of superb quality and immense appeal. Zinc-covered, cedar-lined trunks for tropical use, vellum items and circular, black leathercloth cases stamped 'Sac Chauffeur' especially for use by the motorist, are also rare and unusual. Most pieces of Louis Vuitton luggage of the rigid variety bear a lock number and a unique serial number on the interior label, which will allow the company, always helpful when requested to assist in the research of the history of a piece, to check the archives and confirm the date and place from which an article was supplied. However, they will respect the privacy of the original purchaser by refusing to divulge the identity of whoever commissioned or bought it. Records have been kept since 1854, which enable Louis Vuitton to re-make trays, handles and keys to the original pattern – a legendary service almost unique in the present day.

'Brass-bound' Louis Vuitton trunks – i.e. those with protective corners made of brass, with an edging trim of leather or composition (a tan-coloured synthetic trim developed by the company *c.*1910 in response to the problems caused by real leather proving less durable than all other components) – are more collectable

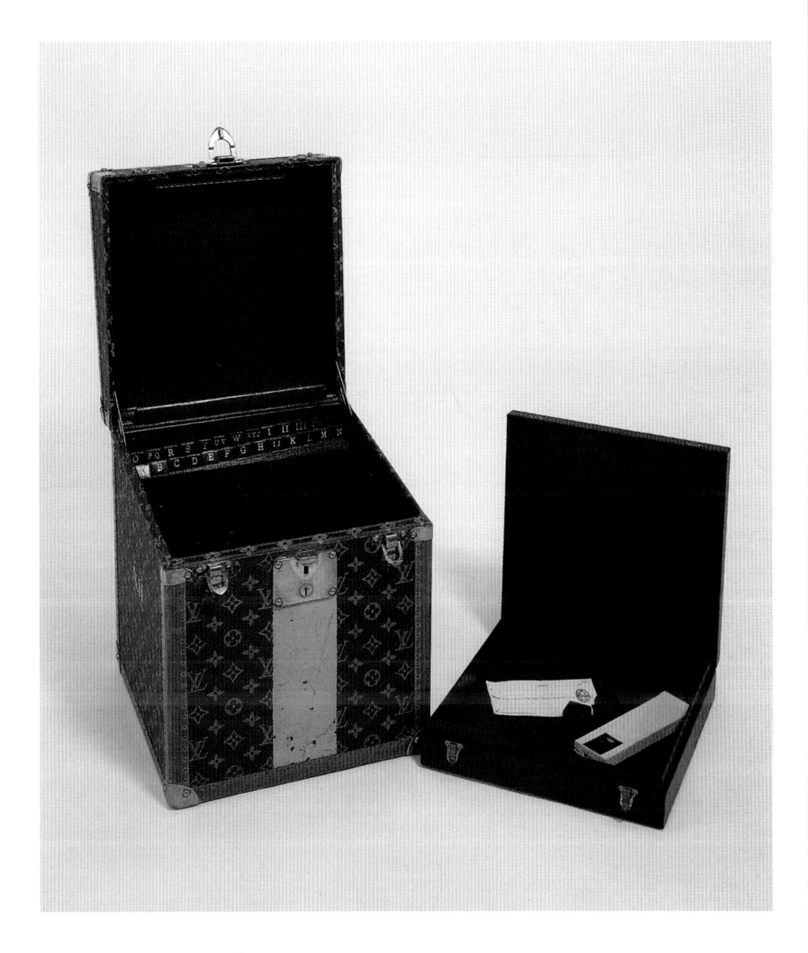

Louis Vuitton monogram-fabric, A-Z file-case. The red box to the right fits snugly into the lid when closed.

Opposite
Drew & Sons 'En-Route' four-person tea-set.

LOUIS VUITTON
LONDON
PARIS
E.B
E.B

than 'iron-bound' examples, which have white-metal protective corners with metal-edging trim, usually painted black. Most examples are adorned with a liberal helping of brass studs, which are almost invariably stamped 'LV', and nearly every case and trunk has characteristic catches and a patented lock marked with the stamped name. However, earlier examples with a squarer, 'pre-patent' central lock will not bear the name of Louis Vuitton on the lock, but will usually retain the printed-paper label to the interior – on earlier examples this is finely engraved, highly ornate, and unlikely to be overlooked. These labels frequently illustrate the most recent medals for excellence awarded to the company at the time an article was produced: Louis Vuitton exhibited successfully at most of the prestigious trade fairs and exhibitions of the time. The ingenious inventions and variety of styles produced by this company are unparalleled, and have almost without exception been imitated by others.

One may also find Louis Vuitton medicine-chests, book-cases, hatboxes, desk and shoe-trunks in a variety of sizes and materials. In the 19th century the company even invented a bed (the Campaign Bed Trunk), which folded up into its own containing trunk. Louis Vuitton luggage is highly coveted, and even modern pieces are of value. The company continues to create special editions and to make developments, whilst also providing its traditional customising service. These special-order items, whatever their age, will always be in demand. Many collectors confine their interest to this one name, not surprisingly given that any piece of Louis Vuitton luggage is likely, if not always to appreciate in value, at least never to depreciate.

Yet another renowned producer of fine leather goods is Hermès (Paris & London), now a household name for its silk-square scarves – produced in dazzling designs and jewelled colours – and for the equally enduring and ever-covetable 'Kelly' handbag popularised by the late Princess Grace of Monaco. Whenever a vintage 'Kelly' bag is offered at auction, it will be hotly contested, whatever the colour and whether of crocodile skin or cowhide. This truly iconic design is still produced by Hermès, but demand continues to exceed supply. Also of great interest to the collector of antique luggage are fitted Hermès dressing-cases containing silver-topped bottles and other accessories, often held in a display tray ready for use, which may additionally fold in half to become a separate case containing only the fittings.

Moving across Europe, the name of Moritz Mädler (an Austrian or German maker), particularly when found on crocodile luggage, gold-stamped onto the front-base section above the handle and under the lid, is very collectable. Moritz Tiller (Austria), another prolific manufacturer of an apparently wider variety of products including early, black leathercloth hatboxes, canvas-covered trunks and wooden-hooped cases with decorative brass studs and fittings, can also be included on the list of great makers.

American companies of note include Hartmann Trunk Company, a prestigious maker of the substantial luggage that is now often used, when it can be found, as furniture. The lovely, durable, mottled canvas, which was often employed for heavyweight wardrobe trunks has a timeless appeal, and its substantial quality has ensured that many items survive in perfect condition.

Opposite
Rare, matched pair of Louis Vuitton leather trunks in exceptional condition and, most unusually, with consecutive serial numbers. Each retains its original webbed linen, lift-out liner (*behind*) and a pair of identical, ornately engraved paper labels pasted to the back and front-interior base (a single label is the norm). The alloy binding is also rare (more common would be an iron, leather or composition trim).

Four Louis Vuitton suitcases in different materials.
Top to bottom
Classic 'LV'-monogrammed canvas; vellum (white leather); black leathercloth (this and preceding case have different types of handle, each recently replaced exactly to original pattern by the company); natural leather.

'Samsonite' and 'Globe-trotter' are American brands still produced today; early examples are collected and cherished. The distinctive, shaped cases in unusual colours produced by Samsonite are particularly interesting to collectors of 1950s American artefacts. Last but not least, Tiffany, the famous American company with branches worldwide, most renowned for glassware and jewellery, also produced exquisite leather goods, and still supplies high-quality leather goods in its U.S. branches.

One could easily devote an entire volume to each of the above names, and inevitably, a summary of this type will omit brands and manufacturers dear to the hearts of other collectors. The Directory of Makers and Retailers, starting on p.89, indicates the scope of further investigation possible, yet is by no means exhaustive. Indeed, one of the thrills of collecting in this relatively uncharted field is the chance of uncovering the mark of a long-forgotten maker or retailer. The piece may even be a sole survivor, providing unique evidence of a manufacturer's existence. Some collectors are excited by the hunt for the obscure, whilst others prefer names of renown. Some choose to embrace the widest possible scope, but many others limit their search to one manufacturer alone, since once one begins to appreciate the merits of vintage luggage, it can become difficult to resist.

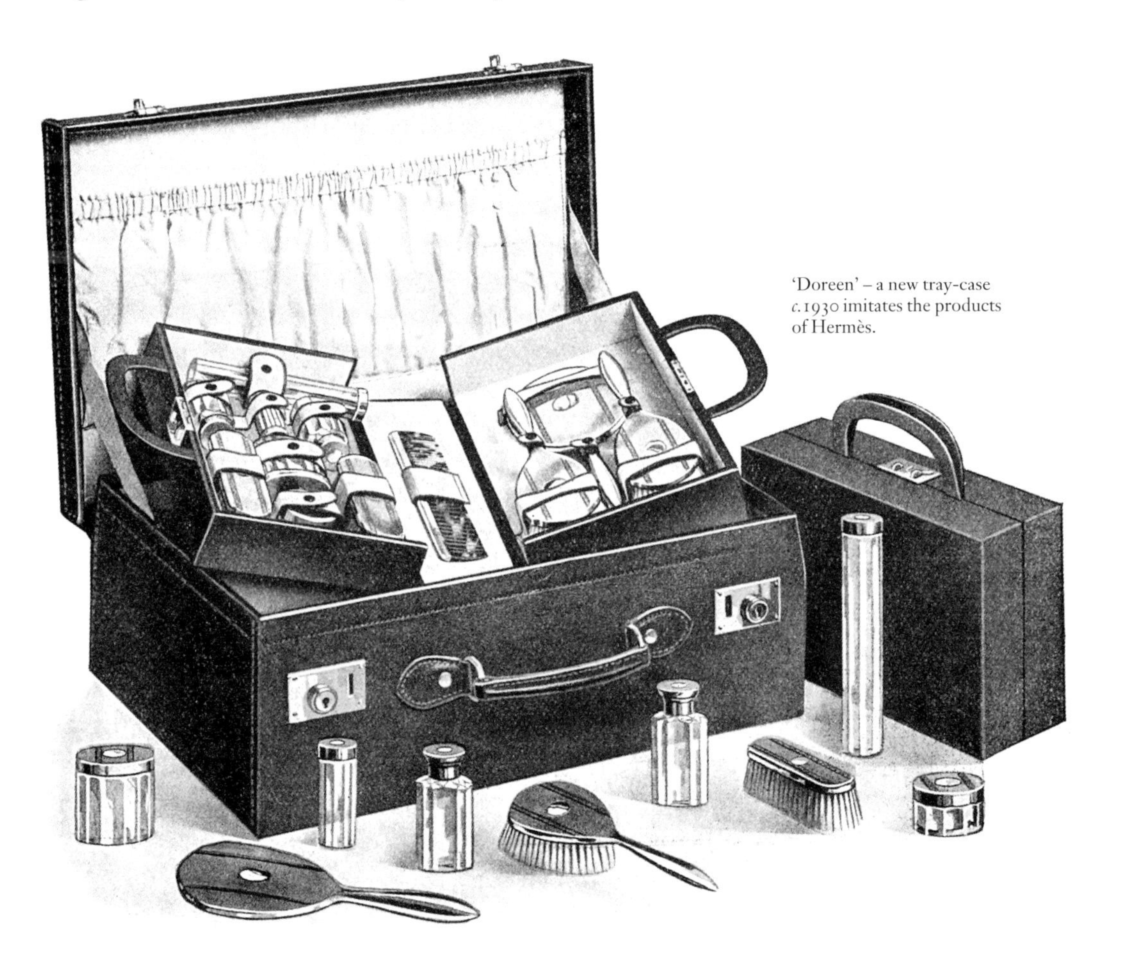

'Doreen' – a new tray-case *c.*1930 imitates the products of Hermès.

6

The Case is Closed

Throughout the years, most, if not all, of the great makers of exquisite luggage have employed precious and exotic materials in the manufacture of luxurious travel goods, if only for a small percentage of items over a short period of time. Many of these materials were obtained from what are now endangered species, and even where exotic animal hides were not used, elephant ivory and the non-sustainable hardwood, ebony, were often utilised for fittings.

Eventually, it became obvious that the wholesale plunder of natural resources could not be allowed to continue, since these materials were, of course, also used for musical instruments, furniture, boxes and diverse objects of all kinds. In 1975, increasing awareness and agreement that controls were urgently needed at an international level to arrest the damage already done to the environment, and to work towards prevention of the extinction of endangered species, led to the United Nations' Convention on International Trade in Endangered Species of Wild Fauna and Flora – known as C.I.T.E.S. Currently subscribed to by over 130 countries, C.I.T.E.S. restricts and controls trade in all products obtained from endangered species. The legislation does not completely prevent responsible ownership or sale of items made from such sources, but is, rightly, stringently enforced, and import or (re)export of almost all such goods will normally require permits and certification. Even today, certain crocodiles and the American alligator may, in specific circumstances, and under strict licensing regulations, be farmed and thereafter used in the production of commercial items, whilst also given strong protection in their wild native habitat (albeit much reduced in area). It is vital to check with the relevant authorities that any such goods of whatever age are

The 'Migrator' ('A wardrobe in a hat box – easily carried in the hand'), Patent No. 310105. A typical example of the ingenious innovations of the 1930s.

From top
Pigskin sextant-case by Alfred Clark *c.*1935; green canvas-cased travelling stove by Asprey; unmarked collarbox *c.*1925; cased set of three bottles for toiletries by Allen's *c.*1880; elephant-hide clutch-bag *c.*1930. The 'mock-croc' suitcase with cap corners *c.*1930 has unusual locks with sliding covers over the keyholes. *Front*: 'magic' wallet in dark-brown-morocco leather; leather-cased travelling iron; mock-crocodile manicure set; notebook; triangular purse, all *c.*1930.

legally permissible, and to establish the correct procedures of export and import.

The rules and regulations pertaining to 'worked "Annex A" specimens acquired before 1 June 1947' have recently been revised, and to some extent relaxed, within the European Community, and there is now a general exemption for their sale. 'Annex A' includes some types of crocodile and ivory – the products of creatures that are still threatened – and the relaxation only applies to antiques that can be proven to fit the criteria. If there is any doubt about the regulations as they apply to a particular object, it is easy to contact the relevant C.I.T.E.S. Management Authority, which will assist in ascertaining the extent and type of documentation that may be required. This is particularly advisable given that the treaty is reviewed and amended on a regular basis.

An important aspect of the C.I.T.E.S. convention is that it provokes widespread, informed awareness of the fragility of threatened creatures and recognition of the validity of protection to ensure their continued existence. The stringent controls and implementation of strict penalties for contravention make it imperative that one responsibly considers the matter of purchase and travel before, rather than after, the fact. A reputable auction house, dealer, or shipping agent may well be prepared to assist with enquiries and paperwork.

The majority of collectors, however, seem to find it a simple matter to build a fine and representative collection of travel goods crafted from sustainable materials that do not fall within the remit of conservation of endangered species – and may often prefer to avoid the complex moral questions, environmental issues and potential complications involved.

Before purchasing an item of luggage, it is also wise to be apprised, as far as possible, of its authenticity, and to bear in mind certain factors that should be considered if its history is unknown. When a piece has been handed down through the generations of a family, an element of folklore will apply, but one can be fairly sure that the basic facts of original owner and age are true. In recent times, however, the growing availability and relatively cheap cost of colour photocopying has led to the easy reproduction of travel labels, and many attempts have been made to 'jolly-up' otherwise mediocre items of luggage with these. It is important to be aware of this possibility when appraising something smothered with colourful labels – differentiating between original printed matter and recent, laser-copied paper is a simple matter. Reproduction labels tend to lie less flat on the surface of hide, and do not adhere as well to canvas as the original labels printed on thinner paper. Spurious labels are also frequently arranged more neatly than those applied in haste just prior to travel. Original labels were usually stuck on with gum, whereas stronger glues are normally needed to make the thicker modern paper adhere. A cheaper trick of the trade is to photocopy black-and-white printed labels and soak them in tea to 'age' them – again, it is a simple visual and tactile matter to make the distinction between an original and a copy.

One should also be a little suspicious of an applied label in the interior of a case: a certain amount of creative 'restoration' might have occurred if the sole sign of intrinsic quality is the little ivory 'Harrods' trade label, or the oval, red-morocco leather, gold-blocked 'Drew & Sons' label. These have been known to be re-applied to inferior products, though thankfully, the practice is relatively rare.

As in all fields of antique collecting, many minor modifications can be made by unscrupulous dealers in order to render a piece more immediately attractive to the unwary. Most collectors prefer antiques that have not been interfered with, and the graceful and unmolested passage through time is the main factor that lends enduring

Three widely travelled cases with hotel labels. Note the thin, stamped, alloy locks on top case, *c.*1930.

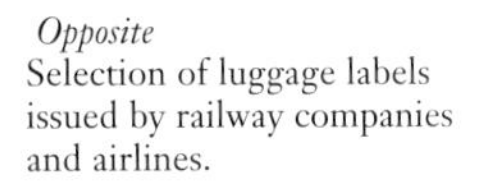

Opposite
Selection of luggage labels issued by railway companies and airlines.

BRITISH RAILWAYS
LUGGAGE IN ADVANCE.
BEXLEYHEATH
4/6
PAID
ONE OF THESE LABELS TO BE AFFIXED TO EACH PACKAGE.
WATERLOO
VIA PORTSMOUTH.
BEA
BRITISH EUROPEAN AIRWAYS
G.W.R.
PADDINGTON
For Transfer to
L. & N. E. R.
[G. E. SECTION]
PREPAID PARCEL
G.W.R.
HENWICK
10d
3
SOUTHERN Rly.
SWANAGE
2d
3/-
Swissair
WITH CARE
Security
PARCEL LABELS
Gummed & Printed ready for use
Fly
TWA
Constellation
LONDON & NORTH EASTERN RAILWAY
NEWCASTLE
CENTRAL
THE QUEEN OF SCOTS
SOUTHERN RAILWAY
VICTORIA
TRANS AUSTRALIA AIRLINES
TAA
LUGGAGE
LMS
LUGGAGE IN ADVANCE
THE QUEEN OF SCOTS
LONDON & NORTH EASTERN RAILWAY
EDINBURGH
WAVERLEY
BRITISH RAILWAYS
LONDON
KINGS CROSS
THE FLYING SCOTSMAN
PLA
PADDINGTON
SABENA
COMPAGNIE INTERNATIONALE DES WAGONS-LITS ET DES GRANDS EXPRESS EUROPÉENS
SIMPLON-ORIENT
EXPRESS
ST. MORITZ
92
LONDON-Vict.
Basel - Laon - Boulogne
BAGAGLIO
403
LONDRES
VICTORIA
TO NEW YORK VIA BRUSSELS
LONDON & NORTH EASTERN RAILWAY
LONDON
KINGS CROSS
Stockholm Central
GARE DE DÉPART
CANNES PLM
467
Fr. Malmö C

appeal to collectable antiques of any type. Realistically, it is still possible to unearth fine examples of antique luggage that have remained untouched for many years but it is more likely that most luggage offered for sale will at least have been given a cursory polish. This is by no means a cause for concern if the work has been carried out sympathetically and with an eye to enhancing the character of the piece, rather than as an inept attempt to 'make perfect' what appear as flaws, rather than patination, to an uneducated eye. Exercise judgement according to your own taste – some people prefer to purchase a fully and expertly restored item, whilst others will only consider a completely original and untouched example.

Those bent on absolute, pristine perfection may be happiest with modern, high-quality travel goods. A great many people, however, prefer to purchase and maintain an antique piece for the simple reason that aged leather has a certain character and dignity similar to that of classic cars or fine furniture. The same people also tend to eschew the plethora of gimmicky offerings available, opting instead for a plain, elegant simplicity of design and style.

One's baggage, after all, becomes an emblem of self-image when travelling. The sense of substance exuded by classic luggage alleviates the anxieties and uncertainties of travel, its reassuring presence representing a certain continuity and security. The appeal of traditional luggage, a solid and reliably familiar companion throughout any adventure, is beyond fad and fashion – speaking to us of home when away, and reminding us of our travels when the journey is complete.

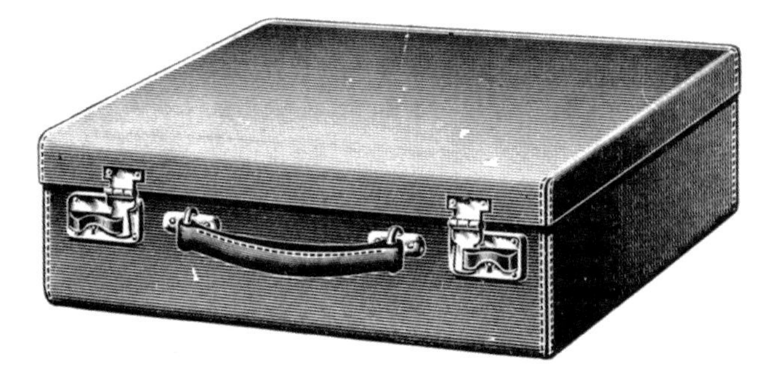

Directory of Makers and Retailers

Abercrombie & Fitch

New York, USA

Gold-blocked maker's name most notably found on rare 'Prohibition cases' of the late 1920s. On the exterior these appeared to be either structured or 'satchel'-style brief-cases but usually concealed three metal bottles for the illicit transportation of alcohol.

'Airport' Luggage

Brand name of lightweight alloy luggage (see illustration on p.13, *top left*) usually spray-painted in various shades of brown when in original condition, bearing a round red, blue and white transfer logo at the mid-point beneath the handle. Dating from the late 1920s to early 1930s they are rare, though underrated at present. Many examples have been 'restored' by stripping the paint down to the metal, which removes any vestige of the applied, exterior maker's mark, and thus the original identity.

The Alexander Clark Co. Ltd

Late 19th-century address:
188 Oxford Street, London W
1930s addresses:
125 & 126 Fenchurch Street, EC3
8 & 9 Fenchurch Court, EC3
38 Leadenhall Street, EC3
1 & 2 Billiter Street, EC3
Works:
98 & 100 Lansdown Road, Dalston E8
Willman Grove, Hackney E8, London
Hylton Street, Birmingham

Relatively prolific company advertising as dressing-case makers, though many fine suitcases exist. These sought-after cases, traditional in appearance, frequently with protective cap corners of leather, were of strong construction and design, rarely needing major repair. Maker's name usually marked directly onto front-exterior rim of the suitcase base, stamped black onto the leather, in distinctive script – applied label never encountered to date. Nickel-on-brass metal fittings usual, occasionally naked brass. (See Mappin & Webb.)

Allen's Portmanteaus

37 Strand, London

Makers of the finest, heavy-duty traditional leather luggage. Appear to have had their heyday from the 1870-90, during which time they held a strong market share of the available trade. Most products were designed for travel by horse-drawn carriage and later by train, and the company, which advertised aggressively until the turn of the century, seems to have faded from view with the advent of the motor-car and demand for more modern luggage. Very collectable indeed – any product certain to be a desirable antique in the true sense of the word. Condition is therefore not quite as important as for those companies trading more recently, but beware of completely decrepit examples, as irreparable damage will inevitably reduce value. Heavy, cast-brass fittings, sometimes nickel-plated.

Antler

Famous brand name, still in production. Early, leather pieces are usually gold-blocked with the word 'Antler', accompanying a depiction of a stag's head.

The Arctic Leather Goods Co. Ltd

Yeaman Shore, Dundee

Advertised in the 1930s as: 'manufacturers leather &c. School satchels, music-cases &c.'

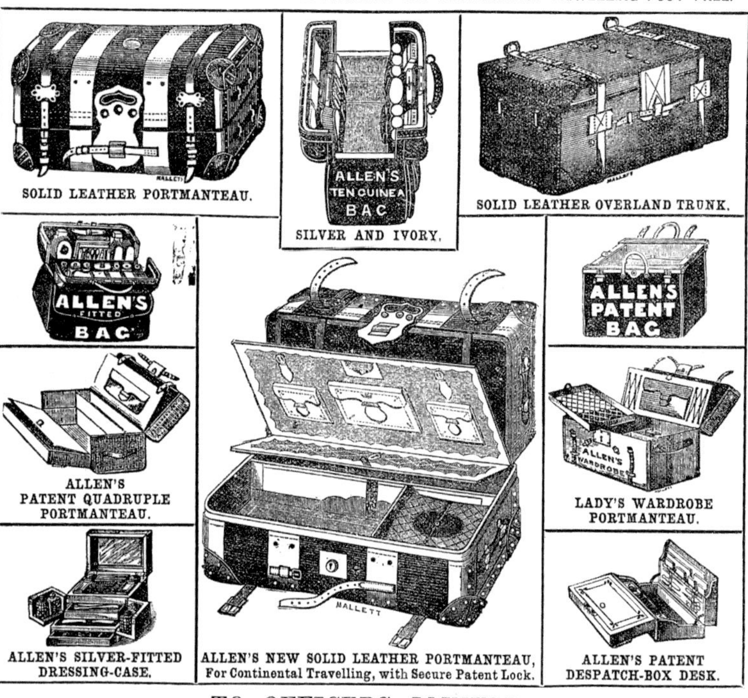

1885 advertisement for Allen's Portmanteaus.

Opposite
Asprey & Co. leather 'Tea for Two' set on top of closed wicker tea-set, Thornhill leather picnic set and Finnigans red morocco-lined tea-set.

Army & Navy Co-operative Society Ltd

105 Victoria Street, Westminster, London SW1

Advertised as dressing-bag makers and trunk and portmanteau makers. Famous military outfitters – still trading today at 101 Victoria Street. Formerly issued a widely distributed catalogue offering a variety of wares, which has become a valuable reference for research and a collectable item in its own right amongst antique dealers. Look out for distinctive cases and hatboxes, some leather, some canvas, leather-bound on lateral edges only. Often marked with a shield-shaped red, green or black morocco trade label, gold-blocked 'A&NCSL', approximately 1 inch square. Additionally, or alternatively, an oval stamp may be directly impressed to the front exterior rim of the item, placed centrally above the handle, only visible when the lid is opened, and about ¾ x ½ inch. Product range extremely varied as to quality, type and age – from bucket-shaped and square canvas and leather hatboxes through leather and canvas (occasionally crocodile) suitcases to trunks of every description. Prolific but desirable name. Brass or nickel-on-brass fittings.

Artistic Bag Co.

28 Redhill Street, London NW1

Traded in the 1930s. Little known.

Ashtona

See Norfolk Hide.

Alfred Askew

99A Charing Cross Road, London WC2

Trunk and portmanteau makers trading in the 1930s. Little else known.

Asprey & Co. Ltd

1915 address:
165 & 166 New Bond Street, W
1930s address:
165–169 New Bond Street, W1;
22 Albemarle Street, W1

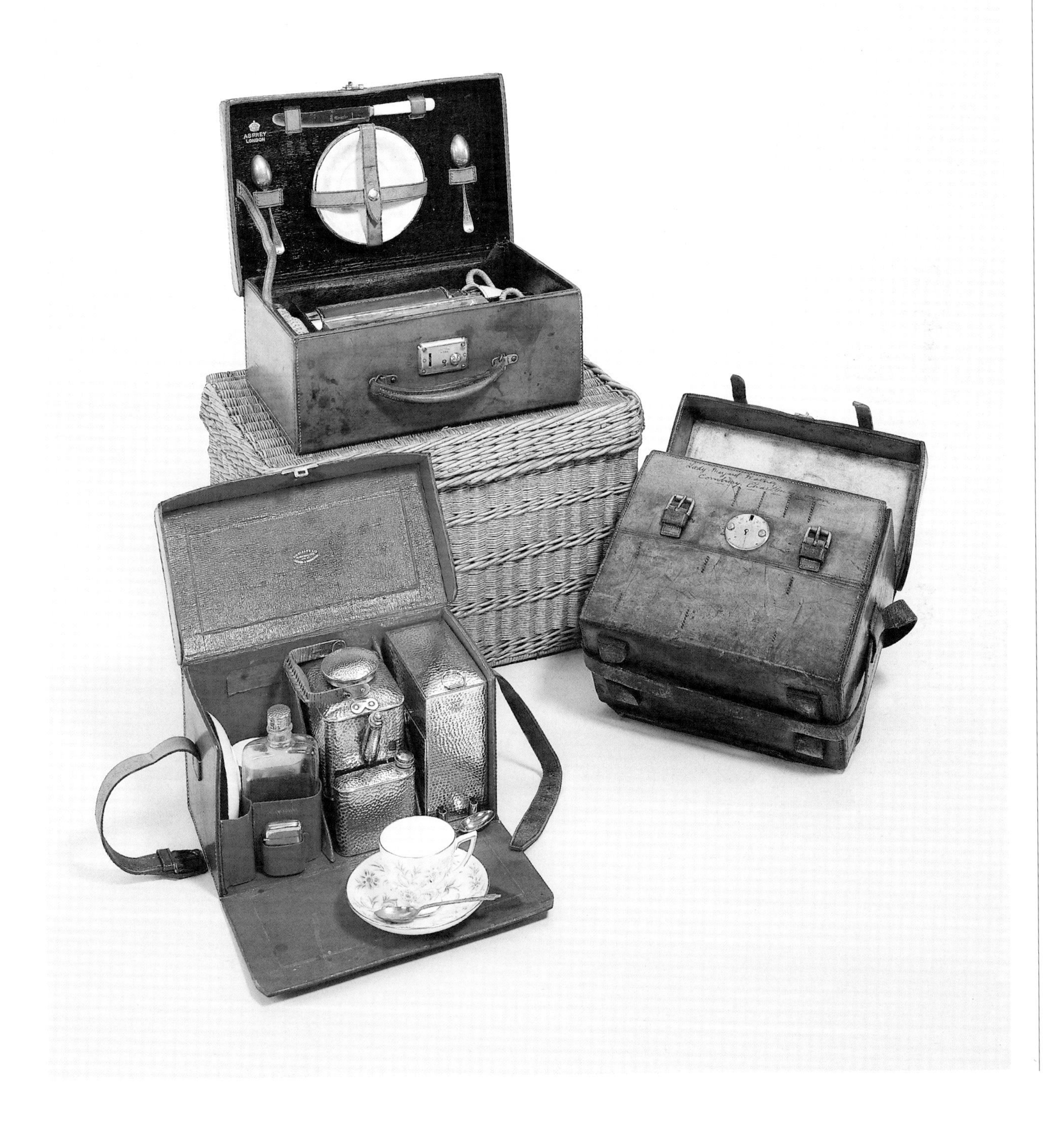
ASPREY
LONDON

Current address:
165–169 New Bond Street, London W1Y OAR

Royal Warrant Holders

Founded in 1781 in Mitcham by William Asprey, the company moved to 166 New Bond Street in 1847. In 1851, Charles Asprey defined his merchandise: 'articles of exclusive design and high quality, whether for personal adornment or personal accompaniment and to endow with richness and beauty the tables and homes of people of refinement and discernment'. Along with objects of many kinds, the tremendously varied range of travel goods produced down the years, from dainty wallets to a magnificent pair of massive, silver-mounted crocodile *necessaires de voyage* (large dressing-cases fitted with every imaginable article likely to be needed whilst travelling) certainly meet the above definition. Invariably producing exquisite leather goods, the company has thrived and made many innovations, absorbing several other firms known for the production of high-quality travel goods in the 19th century.

In 1857, Charles Asprey patented a handle that lay flat when a case was not in use, which became a distinctive feature on many articles. In 1859 the company absorbed Edwards of King Street, Holborn, took over the London branch of Leuchars in 1888, acquired the lease of the property of luggage business S. Last in 1902, and in 1906 bought out the firm of Houghton & Gunn. Awarded a Royal Warrant by Queen Victoria in 1862, and by her successors to this day, Asprey & Co. have consistently produced sought-after goods of the finest quality. It has always been possible to commission unique articles, and the traditional standards of service and quality are still upheld, remaining in the capable hands of members of the Asprey family.

'Astral'

Brand name of mid 20th-century lightweight alloy luggage manufactured by the Heston Aircraft Co. Ltd. (See illustration on p.13, *top middle*.) A metal label attached at the mid-point of the handle gives brand name and serial number, and the tin side-catches may be stamped: 'BRITISH PATENT NOs. 165333/ 173997 MADE'. Originally spray-painted, usually bottle green with two leaf-green stripes around the middle, meeting at the central lock, these cases are now usually found stripped and polished down to bare metal. Later produced in a wide variety of colours. (See Heston Aircraft Co. Ltd.)

Edwin Alex Atkins

8, Shepherd Street, Mayfair, London W1

Advertised in the 1930s as trunk and portmanteau dealer.

Frederick Walter Atkins

269 Camberwell New Road, SE5
County Leather Works, County Grove, Camberwell SE5

Advertised in the 1930s as trunk and Portmanteau makers. Possible family/ trading connection with Edwin Alex Atkins (see previous entry).

J. E. Atkinson

24 Old Bond Street, London W1

Little known. Examples with fine, hand-tooled decoration a speciality, smaller items, such as collarboxes, in particular. Maker's name and address impressed directly onto leather inner rim of the item. Morocco lining probable. Royal Warrant Holders as perfumers, producing, amongst other things, 'Atkinsons Gold Medal Eau de Cologne'.

Attaché Case Manufacturing Co. Ltd

1930s address:
Endurance Works, Blackhorse Lane, Walthamstow, London E17

Along with the Suitcase Manufacturing Co. Ltd, this company came under the umbrella of S. Noton Ltd of the same address and usually the product will be marked 'Noton', the above name appearing to have been used mainly for advertising purposes in directories. Medium quality. Plain, naked-brass fittings, not necessarily cast metal or heavy-weight quality.

Aviation Luggage Co.

Brand name found on mid 20th-century lightweight luggage.

John Bagshaw

Liverpool

Maker of fine, classic leather luggage, similar in appeal and appearance to the products of John Pound. Almost invariably marked with name above left-hand lock, and town above right-hand lock, and stamped in gold along the horizontal upper rim of the base, visible from above when the case is opened. Rare and avidly collected.

The Bag Stores

Northampton

Manufacturers of 'Orient Make' vulcanised-fibre cases and trunks, along with a variety of leather luggage.

Charles George Bailey

51 Trafalgar Road, London SE15

Advertised as trunk and portmanteau makers in the late 1920s/early 1930s. No signed pieces encountered to date.

Thomas & William Bailey Ltd

231–243 St John Street, London EC1

Specialised in surgeons', dentists' and nurses' bags. Since only a small proportion are stamped with a maker's name, it is quite possible that an unmarked example may have started life here.

Bainbridge & Co.

Trunk-makers and Saddlers

Newcastle-on-Tyne

Evidently also a retailer of luggage produced by other manufacturers, since the black-stamped-on-white metal trade label has been found on such items as 'Orient Make' vulcanised-fibre suitcases.

Karl Baisch

Mercedeßtraße 25, Bad Cannstatt, Stuttgart, Germany

Manufacturer of highly prized sets of motoring cases produced specifically for Mercedes-Benz cars.

Henry M. Baker

40 Shawfield Street, Chelsea, London SW3

Advertised as trunk and portmanteau makers. Little known, and no marked pieces encountered to date.

Arthur Barber

Trunk and Suit Case Maker

Broadway, Bradford

Classic, medium-grade suitcases, usually bearing an elongated, oval, paper label, 2 x 1 inches, bordered in red and stamped black with the above details printed red in reverse. This retail label has also been found on 'Revelation' brand cases.

John Frederick Barber

20 Eltham Road, Lee, London SE12

Advertised in the 1930s as leather goods manufacturers.

Albert Barker Ltd

5 New Bond Street, London W1

Royal Warrant Holders (1915) as Silversmiths and Dressing Bag Makers. Still trading up to 1939. Not to be confused with John Barker & Co. Ltd.

John Barker & Co. Ltd

Departmental stores

1930s address:
26–50, 62–70 & 63–97 Kensington High Street, W8
2–16, 13, 19 & 21 Young Street, Kensington W8; King Street, W8
Clarence Mews, W8
2–6, 38 & 39 Kensington Square, W8
Trunk Factory:
134 King Street, Hammersmith W6

This renowned departmental store had its own luggage factory in the early years of the 20th century, and produced a range of luggage of fine quality. A relatively prolifically produced item was a crocodile blouse-case, usually lined with purple or chocolate-brown silk, distinctively 'dumpy' in shape, being slightly deeper than most cases in relation to width and length. The proportions of this attractive style of case are practical for use as a weekend or overnight-case. Gold-blocked 'Barker's Kensington' on the base, above the handle so that the mark lies under the lid when the case is closed. The classic Barker's crocodile case is a particularly collectable item of luggage. The store is still trading in its well-known and imposing premises in Kensington High Street.

Barr of Bristol

Discovered on late 19th-century leather suitcase.

A. Barrett & Sons

63 & 64 Piccadilly, W

One of the most highly regarded producers of traditional English luggage of the finest quality. Held Royal Warrant as 'Brush and Leather Goods Manufacturers, Silversmiths etc' from 1913 (no longer listed as such by 1923). Rare, but frequently well-preserved dressing-cases and picnic sets, usually in top-grade cowhide or crocodile. Fittings are always of finest quality, usually silver. (See illustration of silver-fitted, dome-topped, crocodile picnic set on p.60.) Brass, nickel-on-brass or gilded-brass cast-metal fittings; split-ring handle fixing most unlikely to be original, as cast 'D' ring fixing was standard.

W. Barrett & Son Ltd

9 Old Bond Street, London W1

Advertised as dressing-case makers and suitcase makers in the 1930s. Royal Warrant Holders for Brush and Leathergoods Manufacturers, Goldsmiths and Silversmiths (1948). Possibly a continuation of the family company of A. Barrett & Sons of Piccadilly (see above).

Edward Beaumont

167 Old Kent Road, London SE1

Advertised as leather case maker pre-1939. No marked cases found to date.

Belford Trunk Stores

27 Sicilian Avenue, London WC1

Advertised as trunk and portmanteau makers in late 1920s. Applied trade label and medium quality likely.

Benjamin Bros.

360 Kingsland Road, London E8

Advertised as leather case makers. No marked pieces found to date.

Benjamin, Hollams & Co.

24 Belsham Street, Morning Lane, London E9

Leather case makers specialising in expanding suit and attaché-cases. Also advertised as trunk and portmanteau makers.

Nathaniel Benjamin

3, 39 & 41 Shepherd Street, Piccadilly W1; 1 Clarges Street, Piccadilly W1

Listed in the 1930s as trunk and portmanteau dealers. Very rare, superbly crafted, hand-stitched luggage of the heaviest grade cowhide. Cases marked on the inner-front rim with a metal, stamped label, black-blocked reverse onto white metal, the large capital letters appearing silver on a black ground with address in smaller capitals:

'N BENJAMIN
SHEPHERD MARKET
MAYFAIR, LONDON'

Locks are very large, gilded, heavy, cast brass, stamped 'SECURE LEVER' or 'SECURE 2 LEVER'.

See illustration on p.34 of pair of large, solid-hide, hybrid suitcases/trunks, each equipped with both suitcase handle and pair of side carrying handles typically found on trunks. These provide a thrilling example of the way in which the tastes and choices of people whose existence has been little documented can be brought to life by a few fragments of paper applied to their luggage. Original owner's name boldly stamped on lid:

'J GORDON DUFF
RIFLE BRIGADE'

along with corroborating paper labels and tags giving an address confirming the identity of the owner listed as follows in *Kelly's Handbook to the Titled, Landed and Official Classes 1948*:

> '**Gordon-Duff**, maj. John Beauchamp. M.B.E. (1944); s. of Archibald Hay and Lady Frances Gordon-Duff; b. 1899; educ. Winchester and R.M.C.; m. 1937, Ellen Susan, dau. of hon. Charles Platt Williams, of Lyons, New York; 1 dau.; served in Great War 1914–19 with Rifle Bde, A.D.C. to Viceroy of India (Lord Irwin) 1926–30; A.M.S. to F.M. Viscount Gort in Gibraltar 1942, Malta 1942–44, Palestine 1944–45; D.L. (1945) Aberdeenshire: Naval and Military Club; Cobairdy, Huntly, Aberdeenshire (Forgue 233).'

Other tags and labels provide evidence of extensive travel, which ties in with the above information. Additionally, the accompanying hide suitcase by Paul Romand (Paris) bore a paper label filled out: 'Mrs J. Gordon-Duff' and the initials stamped on each side of this case were – 'E.S.W.'

A paper luggage tag attached to the handle of one of Mr J. Gordon-Duff's cases bears the handwritten words 'Wedding dress (pink stain)'.

We can see that Mrs Gordon-Duff travelled on the ship 'Nederland' of the Nederland Line, Royal Dutch Amsterdam, at some stage, and that the luggage was sent in advance to Folkestone Junction, the 4s 6d charge crossed out and amended to 5/- in blue crayon (it is a very large case!) There are Cunard labels and L.N.E.R. parcels stamps, and we know that the luggage passed via King's Cross, London; Buchanan Street, Glasgow and Waterloo, London at various times.

All these little details, allied to the entry in 'Kelly's', help to confirm and bring alive the history of the example cited above – a delightful coming together of substance and printed ephemera, epitomising the fascination of marrying little details in order to piece together the history of an individual item.

James Benson Ltd

4 & 5 Tottenham Court Road, London W1

Advertised as bag, trunk and portmanteau makers between the wars. The large, red-morocco label, 5 x 2 ½ inches, found in one early example states 'James Benson. Manufacturer of solid leather trunks, bags, suitcases etc. Only address 1, Great Russell St and 3, 4, 5, 263, Tottenham Court Road London'. This was presumably designed to avoid confusion with J. W. Benson Ltd.

J.W. Benson, Ltd

62 & 64 Ludgate Hill, EC4
28 Royal Exchange, EC3
25 Old Bond Street, W1
38 La Belle Sauvage Yard, EC4

Dressing-case makers who also advertised 'Travelling bags with toilet fittings complete'. The surprising rarity of marked examples relative to the company's many premises would indicate that they did not always mark their goods.

E.E. Berry

Cardiff

Classic, medium-grade, cowhide cases, usually with corners. Maker's name will be found stamped directly onto leather of front-exterior under the lid. Locks of naked, cast brass, stamped 'British Make'. Linen-lined.

F. Best & Co.

64A South Audley Street, London W1

Sought-after trunk and portmanteau makers. High-grade, honey-coloured, hand-stitched, cowhide cases with naked-brass locks. By the 1930s the company produced a few leathercloth (usually green), fitted picnic sets. Heavy-duty, cast-brass and (later) nickel-on-brass fittings.

G. Betjemann & Sons

Pentonville Road, London N1

Father of the poet, Sir John Betjeman, Ernest Betjemann, who ran the company in the 1920s and 1930s, was a keen antique collector. He produced many of the fittings for luggage created by Asprey & Co., and Betjemann is credited with the idea of reviving shagreen in the 20th century. Asprey & Co. successfully launched the revived art, making shagreen highly fashionable in the 1920s. The company produced objects and fittings of all kinds from exotic woods and ivory, carrying out work of unparalleled quality and craftsmanship.

John Biffen & Son

77 Albany Street, Regent's Park, London NW1

Trunk and portmanteau dealers between the wars. Probably one of the retail outlets for the many documented manufacturers of which no marked examples seem to survive.

R.A. Blair Ltd

68A Willow Walk, London SE1

Advertised as suitcase maker in the early 1930s. No marked example encountered to date.

A. Boswell

Luggage and leather goods

74 Hanover Street, Edinburgh

Established 1790. One of Scotland's finest and most prolific producers of the highest quality luggage down many generations. Always in excellent condition, weighed against age and use. Eminently collectable and frequently still functional. Usually nickel-on-brass, heavy-quality, cast-metal fittings; naked brass also possible.

H. Boswell & Co. Ltd

1, 2 & 3 Broad Street, Oxford

Usually medium-grade, cowhide suitcases. Generally bear a small, paper label, oval in shape, approximately 1 x 1/2 inch, bordered in red and stamped black with the name and address printed red in reverse, name around the top curve, address around the lower, and the words 'TRUNK MAKERS' across the middle. Larger oval label, as previously described but approximately 1 1/2 x 3/4 inches also discovered on 'Revelation' brand cases. Medium-to-heavy quality, cast-brass or nickel-on-brass fittings.

Brachers

Makers

Bristol & Cardiff

Advertised as bag, trunk and sample-case manufacturers. One of the most highly regarded of the West Country and Welsh makers of top-quality cowhide luggage. Heyday was the late Victorian/Edwardian era. Brachers cases have distinctive, heavy-duty and superbly crafted handles, very similar in appearance to those of Insall, with whom their cases share a design flaw: they were made without a 'keep', which would have prevented the user from accidentally bending the lid section right back and tearing it from the case. It is therefore very rare indeed to come across a case by Brachers in entirely original condition – they are likely to have an old 'strip' repair to the back seam, which has usually mellowed to an acceptable match with the original hide. Most examples are unbleached linen-lined, with a leather-edged lid flap, accompanying leather straps, a pair of retaining straps in the base and an additional gold-blocked, leather trade label, on tan hide or red morocco, usually, but not always, oval in shape. Rarely, the lining is of caramel hide, probably stamped to imitate crocodile. Maker's name is also invariably found impressed on each lateral side of the lid. Always large, cast-brass, or more probably nickel-on-brass locks, and always large 'D' ring (not split-ring) handle fixing.

Bramah & Co. (Needs & Co.)

1874–1901 address:
100 New Bond Street, London W
Current address:
Bramah Security Equipment Ltd
31 Oldbury Place, London W1M 3AP
Tel: +44 (0) 171 486 1739

Suitcases, produced in their entirety by the company – as distinct from the metal fittings widely supplied by Bramah to many leather goods manufacturers (marked 'Bramah') – bore an impressed mark on the body of the leather on the front-inside rim stating:

'BRAMAH & Co.
Makers to the king
100, NEW BOND ST. W'

and the naked-brass lock is engraved:

'JT Needs
100 NEW BOND St'

along with:

'late [engraved crown logo]
J BRAMAH
124 PICCADILLY'.

The current product catalogue states:

> 'All old Bramah locks are serviced by Bramah Security Centres Ltd and where possible repaired, opened and new keys cut. This is a specialised service using historic and occasionally temperamental equipment. We will always quote for work in this area but it does take time.'

At present this will usually amount to less than £20 if a copy of a key in your possession is required, and about double otherwise.

(See table A on p.135 for chronology of Bramah locks.)

R. Branker & Son

42 St John Street, London EC1

Obscure dressing-case maker of the early 1930s. It is possible that marked examples exist; if so, they are very rare.

W.S. Brewer & Son

Manufacturers

120 Oxford Street, London W

A gold-blocked, dark morocco-leather label, rectangular, with canted corners, applied to the interior lid identifies cases produced by this early 20th-century maker.

'Brexton'

Popular brand name of mid 20th-century picnic sets. Although not as appealing to the purist antique collector as the earlier products of such companies as Drew & Sons, G.W. Scott etc., these later picnic sets of the 1950s and 1960s, often with Bandalasta (a type of plastic) or other plastic fittings, are becoming increasingly collectable, particularly amongst classic-car enthusiasts owning a vehicle from the period.

W.W. Bridge

23 & 24 Wormwood Street, Old Broad Street, London EC

Advertised as portmanteau, bag, saddle and harness makers. Classic, hand-stitched ('saddler-made') luggage. Some very fine, large suit-cases and leather trunks survive, marked 'W.W. Bridge' or 'Bridge Portmanteau', either impressed onto inner rim of hide body or, in addition, on morocco-leather label. Excellent quality but not particularly sought after – nonetheless well worth consideration if in good condition. Heavy, cast-brass, or nickel-on-brass fittings.

Thomas Brigg & Sons Ltd

1915 address: 23 St James's Street, London; Avenue de L'Opera, Paris
1930s address: 49 Rue du Faubourg, St Honore, Paris

Royal Warrant Holders (1915) as stick, whip, umbrella and parasol manufacturers. Advertised in the late 1930s as providing 'Ladies and gentleman's umbrellas, walking stick, hunting crop, riding & driving whip & patent perfect sporting seat manufacturers, by appointment to HM the King, HM the Queen, HRH the Prince of Wales & Royal Family'. Merged shortly afterwards with Swayne & Adeney to become Swayne Adeney Brigg. (See below.)

C. & R. Brinsley

1A Heber Road, East Dulwich, London SE22

Listed as attaché-case makers before the Second World War. Little known.

Robert P. Bristow

21 Great Ormond Street, London WC1

Recorded as dressing-bag makers but marked examples rare.

William Brock & Co. Ltd

174 Weston Street, London SE1

Recorded as leather-bag maker but no further information or marked examples encountered to date.

George William Brooks

Repairer

22A Launceston Place, London W8

Recorded as trunk and portmanteau maker and repairer (probably a large part of his trade). There do not appear to be any marked examples of his product, however. Not to be confused with Brooks Motoring Luggage.

J. B. Brooks & Co.

Motor Car and General

Criterion Works, Great Charles Street, Birmingham 3

Advertising as trunk and portmanteau makers in the first half of the century, this company is well-known for the distinctive, black leather-cloth motoring-cases marked with its name. Often these were produced in sets, which slid like drawers into a larger trunk. Nickel-on-brass, cast locks and clips. Motoring picnic sets of excellent quality were also produced by Brooks.

Alex Brown

88 Moorgate, London EC2

Trunk and portmanteau maker. Nothing else known to date, and no marked examples discovered.

Brown, Best & Co.

44 Tarn Street, New Kent Road, London SE1

Advertised as makers of attaché-cases, leather suitcases, bag, trunks and portmanteaux. No marked examples or further information available.

Cyril Brown

10 Sussex Place, London SW7

Listed as trunk and portmanteau makers. No marked examples or further information available.

Bruss & Co.

59 Stanhope Street, London NW1

Listed as wholesale bag maker. One of many companies operating up to the mid 1930s that supplied unmarked products, or products labelled with the retailer's trade label.

Robert Bryant Ltd

Sharsted Works, Sharsted Street, London SE17

Advertised as makers of motor-touring trunks and golf-bags up to the late 1930s, possibly later. Attractive leather golf-bags relatively commonplace, bearing an oval ivorene trade label with the words 'a BRYANT Product'.

David Bryce & Son

Glasgow

Bryce's produced many of the miniature books – guides to etiquette and dictionaries etc. – that were often included within fitted dressing-cases, frequently bound to match the other leather-covered items such as manicure folders, inkwells and vestas.

'Burne'

Guaranteed plywood throughout

Brand name found on typical, brown-canvas-covered, woodbanded trunk from the first third of the 20th century. Applied trade label stamped black on vegetaline cartouche. (Vegetaline was a type of imitation ivory made by treating wood fibre with sulphuric acid.)

Caplin Harris

4 Ossulston Street, London NW1

Trunk and portmanteau dealers.

Carmichael's of Hull

Distinctive Luggage

The above name and slogan is very occasionally found on a fabric label, woven in white, orange and brown and measuring 2¾ x 1¾ inches, attached to the interior lining of classic, early-to-mid 20th-century suitcases.

H. J. Cave & Sons

81 New Cavendish Street, London W1

Royal Warrant Holders from 1913–15. Advertised as trunk and portmanteau manufactures. Heavy-duty shirt-trunks with pull-out drawers were a speciality of this company, which was no longer listed by the late 1920s.

R. Chadwick Ltd

30 Tyrrel Street, Bradford, England

Early-to mid-20th century. Rectangular, paper label to interior lid, printed brown on ecru, states:

'Makers of suit & attache-cases
Handbags * Umbrellas'

Medium-to-very-good quality. Expect heavy, cast nickel-on-brass locks.

M. Chapman, Son & Co. Ltd

2 Charterhouse Buildings, London EC1

Listed as dressing-case makers.

J. Chaumet

154 New Bond Street, London; 12 Place Vendome, Paris, France

Royal Warrant Holders as jewellers, later producing a range of luggage marked with their name, either with gold-blocking or metal trade label. Sometimes marked 'J Chaumet et Fils Lausanne'.

Cheney

The name of this prolific lock maker is often found stamped on locks, attached usually to medium-grade luggage.

Chubb & Sons Ltd

Pre-1913 address:
47 St Pauls Chyd, London
c.1913 address:
128 Queen Victoria Street, EC

Chubb were Royal Warrant Holders and lock and safe makers in the early years of the 20th century. It is difficult to ascertain whether they also supplied luggage. Many suitcase locks are stamped 'CHUBB'. Earlier examples occasionally bear an engraved chubb fish. Heaviest duty, cast-brass fittings.

E. J. Churchill

Gunmakers

32 Orange Street, Leicester Square, London WC2

Curiously, the examples of gun-boxes available in the present day marked with this maker's name seem exclusively to be for small guns. Usually they have no brass reinforcement corners, and their small size tends to render them less commercially attractive. Not very prolific.

The Alexander Clark Co. Ltd

See under Alexander.

Clark & Son

22 Gloucester Road, London SW7

Trunk and portmanteau dealers, 1930s.

Alfred Clark

33 New Bond Street, W1

Royal warrant holder (1913). Silversmith and dressing-bag maker to the Royal Family in the early years of the 20th century. Few examples extant. One pigskin sextant-case (see illustration p.85) is marked:

'Alfred Clark
1928 Ltd
7 Vigo Street'

Gold-blocked onto front rim, with owner's name 'Lord Louis Mountbatten' gold-blocked to interior lid, suggesting that the company changed address by the late 1920s.

Frederick C. Clark

2A Sternhall Lane, Peckham, London SE15

Advertised as an attaché-case maker in the 1930s. No marked examples encountered to date.

A. R. Clarke Ltd

94 Southwark Street, London SE1

Advertised as suppliers of treated canvas for trunk makers in the 1930s.

S. Clarke & Co. Ltd

33 Bowling Green Lane, EC1
9 & 11 St James' Walk, London EC1

Advertised as trunk and portmanteau makers in the 1930s. Occasional examples. Note the proximity of their address in Bowling Green Lane to Pukka Luggage.

T. Cleaves & Son

Coventry

Found stamped on simple cowhide despatch-bag with strap.

Cleghorn

104 George Street, Edinburgh

Prolific maker of fabulous quality luggage in cowhide, but also notable for leather-trimmed canvas baggage in various shades of grey and green, and superb, specially commissioned pieces. Substantial luggage, usually well kept, and seldom in need of repair. Along with Reid & Todd of Glasgow, appears to have been the obvious choice for quality luggage in Scotland for generations.

Jabez Cliff & Co. Ltd

1930s details:
11 Savoy House, 115 Strand, London
Globe Works, Lower Forster Street, Walsall
Current tel: +44 (0) 1922 621676

Established in 1793 and still based in Forster Street in Walsall, run by family members. Reputedly the oldest golf-bag manufacturer in the world, the company also produces a range of leather, and canvas-and-leather luggage called the '1793 Range' as well as other classic luggage for well-known names. Like many companies, they had a retail outlet in central London along with a manufactory based elsewhere. Walsall is traditionally one of the main centres for leather work.

Jacob Cohen

218 Whitechapel Road, London E1

Advertised as trunk and portmanteau makers in the 1930s. Little known.

Jacob Cohen

172 & 174 Hackney Road, London E2

Listed in Trades Directory, separately from the previous entry in this list as suitcase makers in the 1930s.

Cole Brothers

24A Floral Street, Covent Garden, London WC2

Advertised in the early 1930s as makers of leather cases, trunks and portmanteaux, dressing-bags and bags. Few examples, or at least few marked ones, seem to have survived.

Richard Collins

18 Prince Regent Lane, Plaistow, London E13

Trunk and portmanteau dealers in the late 1920s.

'Coracle'

Trademark of picnic sets produced by G.W. Scott & Co. (See below.)

W. S. Cowell Ltd

Ipswich

The rectangular, midnight-blue paper label, 2 x 3 inches, printed white, indicates that W. S. Cowell Ltd supplied 'travelling trunks and bags – repairs executed'.

'Crescent'

Brand name for expanding suitcase similar in construction to the 'Revelation'. Examples are invariably embellished with naked-brass fittings, and usually the body of the suitcase is of a rich, reddish-brown hue. Attractive and practical, but not as highly regarded as 'Revelation' products.

Albert Cross

35 Fulham Road, SW3

Listed as trunk and portmanteau makers in the 1930s. Not connected with following listing according to current available evidence.

Mark Cross

London

Fairly prolific, popular manufacturer, whose goods are often impressed with a small (approximately 1/2 x 3/8 inch) gold-blocked stamp showing a tiny, leafy logo above the larger word 'CROSS' in curly upper-case script, and a smaller, plainer, upper-case 'LONDON' below. Many examples survive, and are moderately sought after, some in crocodile hide.

Mark Cross was originally founded in the USA in 1845 by Henry W. Cross. In Alfred Hitchcock's film *Rear Window* (1954) Grace Kelly opens a case from this company with the words: 'It's a Mark Cross attaché – compact, but ample enough'. The company has recently updated their range and continues to produce elegant luggage.

'Crown' Brand

Registered trademark of British tin trunks made in the early 20th century, usually painted with brown, scumbled exterior and bright-blue interior.

George Sanderson Crowther

63 Long Lane, Bermondsey, London SE1

Advertised as leather case makers in the late 1930s. No further information presently available.

'Cruzer'

Fortnum & Mason Ltd

Brand name found on vellum luggage.

Dalton & Young

Brokers

28 Fenchurch Street, EC3

Advertised as brokers of reptile skin in the 1930s.

H. F. Daltrey & Co. Ltd

17 & 19 Wagner Street, SE15

Listed as dressing-case makers in the 1930s. No marked examples discovered to date.

August Dams

18 Tachbrook Street, London SW1

Advertised as an attaché-case maker. Despite the relatively central address, which would imply a successful business, no marked examples discovered to date.

Darnborough & Sons Ltd

Fancy Leather Goods

53 Linthorpe Road, Middlesborough, England

Gold-stamped, oval, midnight-blue, paper retail label, 1 x 1½ inches, applied to central interior lining of usually medium to good-quality cases.

Harry Edward Davies

1 Marlborough Road, London SE1

Advertised as attaché-case maker in the 1930s. No further information presently available.

Davis & Co.

10 Strand, London WC

Royal Warrant Holders as Trunk Makers to the Royal Family in 1915. However, the company does not appear to be listed by the mid 1920s. Possible continuation by A. Davis (Piccadilly) Ltd (see below).

A. Davis (Piccadilly) Ltd

Piccadilly, London W1

Advertised as trunk and portmanteau makers, bag and dressing-case makers in the mid 1930s, and although it has not been proved, it is possible that the company was, in some form, a continuation of Davis & Co. of the Strand (see above). No more than a few dozen examples seem to have survived, almost always standard hide suitcases of excellent quality, exquisitely patinated, with cast-brass lock fittings.

D. Davis & Co.

7 Green's End, Woolwich, London SE18

Listed as trunk and portmanteau makers in the early 1930s. No other information, and no marked examples apparent to date.

Day & Son

353 & 378 Strand, London

Early-to-mid Victorian makers of 'Trunk & travelling equipage, camp & barrack portable furniture – MANUFACTURERS – officers, and travellers equipped for all parts of the world'. Likely to have a Bramah lock and to bear a pasted-on, interior, paper-printed, black-on-white label giving the above address and details. Exquisite hand-tooled decoration on heavy bridle leather.

Henry David Dean

145 Edgware Road, London W2

Trunk and portmanteau dealer in the early 1930s. Probable connection with Thomas Walter Dean (see below).

Thomas Walter Dean

30 Caroline Street, Pimlico, London SW1

Trunk and portmanteau makers in the early 1930s. Probable connection with Henry David Dean (see above).

Arthur William Dear

37 Knightsbridge, London SW1

Trunk and portmanteau maker. Heavy cowhide, stamped 'A.W. Dear', black-embossed onto the leather. Few examples exist, usually linen lined.

G.T. Dearberg & Sons Ltd

23 & 24 Charles Square, City Road, London N1

Bag, dressing and writing-case maker.

Debenham & Freebody

1930s address: 40 Wigmore Street, W1
1948 address: 17–37 Wigmore Street, London

Well-known department store, listed in

1930s trades directory as trunk and portmanteau dealers. Trunks or cases with this trade label, or an impressed mark, are found occasionally. The company was probably not the actual manufacturer, possibly contracting out work to some of the many small businesses listed here, about whom little or nothing is known, to produce goods bearing their trade label.

John S. Deed & Sons Ltd

91 New Oxford Street, London W1

Reptile-skin merchants.

Malle Delion

Brevette S.G. D.G.

24 Bvd de Capucines et 15 à 25 Passage Jouffroy, Paris

Above label found on metal tag to the exterior of rare, early 20th-century articles of luggage similar in design and execution (canvas on wooden base) to the products of Louis Vuitton.

Denty Luggage

78 & 80 Edgware Road

Established London 1820. Paper trade label, orange and white on a sky-blue background, edged with brown, occasionally found in cases from the 1920s and 1930s. No information regarding previous history.

Derry & Toms

Kensington High Street, London

The applied trade label of this famous store is found (though rarely) on leather luggage, and occasionally on small wooden cases with fretted white-metal trim and leather handle. The label applied to the latter is approximately 1¾ x 1¼ inches, black-stamped on ivory or ivorene, and includes the word 'patentee'.

Edwin Alfred Dilnutt

6 Chitty Street, Fitzroy Square, London W1

Leather case makers, 1930s

Direct Handle Co.

Fittings

1A Fox's Lane, Hackney, London E9

Advertised as suppliers of suitcase makers' fittings.

Direct Suit Case Supplies Ltd

42 Bethnal Green Road, London E1

Suppliers to suitcase makers.

Dixon & Co.

4 Church Street, Kensington, London W8

Trunk and portmanteau makers.

Dobson & Robinson

16 Wells Road, Oxley

Employed the ubiquitous red and black, circular, paper trade label so popular throughout the 1930s to indicate the origin of the leather cases they supplied. This label lists a saddlery and a sports depot in addition to a travelling goods department.

Edward Doherty & Sons

700–710 Seven Sisters Road, Tottenham, London N15

Advertised as manufacturing 'Dental & surgical bags, nurses' & surgeons' bags... Wholesale'.

A.D. Dormer & Son

Satchels

55 Fann Street, London EC2

Advertised as wholesale satchel maker.

A.L. Downes

40 St John Street, London EC1

Dressing-case makers.

Drew & Sons Ltd

33, 35 & 37 Piccadilly Circus W1
215 Piccadilly W1
156 & 157 Leadenhall Street, EC
Works: 62 Hatton Garden EC1, London

Established in 1844. One of a handful of truly legendary English luggage makers, very much collected and cherished. Advertised as trunk and portmanteau, dressing-case and bag makers, and many fine examples of their luggage are still in circulation. Their innovations in travelling equipment, dressing and travelling-bags were shown at the Inventions Exhibition of 1885. Also famed for picnic sets of all types, chiefly the wicker-cased Drew & Sons 'En-Route' (see illustration on p.78), much sought after amongst veteran car owners. Some time after 1914, moved to 33 Regent Street SW1 and became a limited liability company (no. 229,271) on 30 March 1928.

W. Duke

Travel and Sports Outfitter
19 & 21 Bridge Street, Northampton

Goods supplied by this retail outlet were usually marked with a label, ¾ x 2 inches, printed black on gold paper to the interior lid section.

Henry Duligall & Son

17, Seymour Street, Euston Square, London NW1

Early 20th-century leather case manufacturers.

Alfred Dunhill Ltd

Current Head Office:
27 Knightsbridge London SW1

Tel: +44 (0) 171 838 8000
Retail Shop: 30 Duke Street, London SW1
Tel: +44 (0) 171 290 8600

Boasts a long tradition of producing a variety of fine-quality luggage. The company was named Dunhill Motorities in the early years of the 20th century, and produced several ranges of motoring luggage, including its famed sets of 'Kennard' valises. Such vintage luggage is extremely rare and sought after. The company maintains a comprehensive archive and still supplies fine leather goods.

The Alfred Dunhill Museum at 48 Jermyn Street, London SW1 displays fine examples of luggage. The company considers re-purchase of suitable vintage pieces.

Percy Dunker

131 St George Street, London E1

Advertised as an attaché-case maker in the pre-war years.

Dupont

Rue Dieu, Paris, France

Magnificent dressing-cases and bags, often of Havana crocodile, with Baccarat crystal, vermeil-topped bottles.

Eagle Lock Co.

Made in USA

Terryville, Conn.

Found marked on tin lock of lady's large, round, black leathercloth hatbox with tan leather trim and handle *c.*1910 and on similar locks on various cases that apparently date from the second decade of the 20th century.

Sydney Charles Ebert

19 Featherstone Street, London EC1

Leather case makers, 1920s/30s.

'Eclipse, Regd.'

Gold-blocked brand name found on some cases of the 1920s/30s, particularly vellum items.

Edwards

King Street, Holborn, London

1817–1859 (absorbed by Asprey's in 1859). Leading manufacturers of patented writing and dressing-cases, mainly military. Appointed Dressing Case Maker to William IV in 1832, the Royal Family and the East India Company. Awarded Gold Medal in the Great Exhibition for a gentleman's dressing-case, of interest because it was an innovative transitional piece between wood and leather, being made of wood, like earlier coaching pieces, but the first to be covered in leather (i.e. 'wood-lined'). Earlier pieces tended to be either of unstructured leather, or else held within a plain wooden carcass. Thomas Jeyes Edwards was thereafter awarded the Royal Warrant as Dressing Case, Travelling Bag and Writing Case Maker to Queen Victoria and remained with Asprey's until his retirement in 1872.

Edwards & Sons (of Regent Street) Ltd

161 & 159 Regent Street, London W1

Royal warrant holders in 1913 as stationers, silversmiths and dressing-bag manufacturers. In 1914/1915 'antique dealers' was added to the list. Rare and exquisite crocodile dressing-cases and bags occasionally found.

Eldrid, Ottoway & Co. Ltd

Wholesale

42, 44, 46 & 48 Whitecross Street, EC1
1, 2 & 3 Silk Street, Cripplegate, EC2

Advertised in first third of the 20th century as leather case makers.

Ellenger & Co.

Portmanteau Makers

1 Grey Street, Newcastle On Tyne

Oval, gold-blocked, red-morocco retail label, approximately 3 x 2 inches, found on interior lid of square hatbox *c.*1890 with blue-on-cream striped lining and front handle. The name is also found stamped on each lateral side of the lid.

Empire Trunk & Basket Works

42 Shepherd's Bush Road, London W6

Listed in Trades Directory of the 1930s as basket maker. Probable manufacturer of some of the early lightweight wicker luggage.

Walter English

3, 4, 8, 9, 10, 11 & 12 Royal Opera Arcade, London SW1

Trunk and portmanteau dealers.

Erskine & Sons Ltd

Makers

63 Ann Street, Belfast

Late 19th-/early 20th-century makers of exquisite traditional dressing-cases etc., often morocco lined. The exterior of the case is likely to be decorated with subtle, hand-tooled decoration.

William (William Corrie) Evans

63 Pall Mall, London SW

Gun and rifle maker still trading at 67a St James's Street, London SW1. Superb-quality cartridge and gun-cases of oak-lined cowhide with brass reinforcement corners.

Excelsior Fibre Co. Ltd

140 Canonbury Road
Head office & works: Bacup, Lancs

Advertised as attaché-case, suitcase,

trunk & portmanteau makers. Few marked examples. Probable source of many of the unmarked examples of fibre luggage that are still in existence and use today.

Farringdon Leather Goods Manufacturers Ltd

16–20 Farringdon Avenue, London EC4

Advertised in early 20th century as dressing-case makers.

Arnold Farthrop

263 Kennington Road, London SE11

London agent in the 1920s/30s for Soutar, Laird & Co. Ltd ('mfrs of waterproof canvas, school satchels, market bags, workmen's bags, tool rolls &c.')

'Featherlite'

This brand name, along with a distinctive logo depicting a feather, may frequently be found embossed on a thin, brass, oval label, about 1 x 1⅓ inches long, to each lateral lid section of canvas-covered cases and hatboxes on flax-fibre foundation. A similar pair of labels, also of thin, stamped brass, but triangular in shape, with a picture of a flying stork bearing a light bundle suspended from its beak, may also be found on similar cases.

Joseph Fellner

12 & 12A Roscoe Street, London EC1

Early 20th-century trunk and portmanteau makers.

William Ferguson

60 Kinnerton Street, London SW1

Trunk and portmanteau makers.

F.O. Fesche

2 Carlisle Buildings, Eastbourne

A small, navy-blue, paper label gives the additional information: 'For value in trunks, bags, fancy leather goods and stationery'. Early 20th century.

'The FIACO trunk'

Brand name found on small, stamped-brass label applied to the exterior of brown canvas, wooden-hooped trunk *c.*1925.

Fibre Case Co. Ltd

36 & 38 Peckham Road, Camberwell, London SE5

Suitcase maker of the early 1930s, specialising in lightweight fibre luggage. Few marked examples.

Fichet

France

Produced luggage in imitation of monogrammed Louis Vuitton luggage. From a distance it is quite convincing. On closer examination, however, it becomes obvious that the monogram is in fact a rather coarse rendition of Vuitton's yellow-on-brown canvas theme, composed of three yellow, lobed dots resembling a trefoil or *fleur-de-lys*. This early 'knock-off' was produced in moderately prolific quantities, and, like its modern counterparts, does not bear comparison with the genuine article.

Finnigans Limited

17, 18, 19 & 20 New Bond Street, W1
12 Clifford Street, W1
8 Coach & Horses Yard, Old Burlington Street, W1
Works: 6–10 Lexington Street, W1
Deansgate, Manchester

Ranks amongst the finest of English makers. Advertised over the years as 'Actual makers of best quality trunks', portmanteaux, dressing-bags and cases. Produced many and varied examples of exquisite luggage, and survived until around the late 1970s/early 1980s. The maker's name is almost invariably found impressed on each exterior side of the lid, together with 'London', 'Manchester' or 'Liverpool' or various combinations of the three locations. Early Victorian examples impressed 'D Finnigan, Maker, Manchester' are very rare indeed.

Fisher

188 Strand, London
Est. 1838

Renowned dressing-case maker of the Victorian era. Examples are rare and highly collectable.

Fitzroy Leather Works

56 Grafton Street, Tottenham Court Road, London W1

Dressing-case makers in the 1930s. No marked examples. This must have been a strictly wholesale manufactory.

Flashman, F. W.

Repairer

70 Dulwich Village, London SE1

Trunk and portmanteau maker and repairer in the early 20th century.

Alexander Forbes

73 Fairfax Road, Hampstead, London NW6

Trunk and portmanteau makers in the early decades of the 20th century. No marked examples encountered to date.

1938 advertisement for Finnigans luggage.

★ *Distinguished Leathercraft*

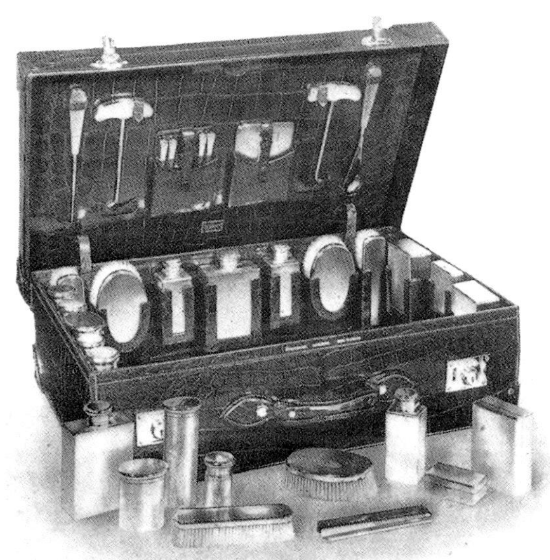

by FINNIGANS

EVERYBODY travels nowadays. What was formerly the prerogative of the more fortunate or advantageous few has now become a social habit. It is an odd thing, however, that while most travellers are rightly fussy about dress for the journey, few seem to care at all about the clothes they take with them. The correct luggage is, of course, equally important, and Finnigans cater for travellers by air, rail, boat, road, or even safari!

★ **Gent's Cedar Crocodile Dressing Case.** Lined Cedar Crocodile and fitted solid Silver, Engine Turned Barley design Fittings, complete with cover. Size 28 inches by 18 inches by 8 inches. **£275.0.0**

★ **Finest quality Raw Hide Wardrobe Trunk.** Chromium plated fittings throughout. Fitted ten hangers, Patent Self-Locking Turn Lock acting on body of Trunk or on Drawers; five Drawers and Shoe Box with Dust Curtain. Lined washable rexine. Finest finish throughout **£29.10**

FINNIGANS 17-20, New Bond Street, London, W.1

Forsyth

Edinburgh

Rare and sought after, the products sold by this renowned departmental store are almost invariably of cowhide, with a distinctive twist: the rim at the base of the suitcase is trimmed with crocodile-effect hide. Usually with heavy, nickel-plated, cast-brass fittings. Name generally black- or gold-blocked in distinctive 'signature'-style script onto the exterior upper-base section. Some early examples are stamped 'R.W. Forsyth, Edinburgh & Glasgow'.

Fortnum & Mason Ltd

1915 address: 181–183 Piccadilly, W1
1930s address: 181–184 Piccadilly

Although primarily known as purveyors of a variety of provisions such as tea and coffee under Royal Warrant, Fortnum & Mason Ltd manufactured and retailed camp equipment and exported goods of all kinds throughout the world, and still do. Occasional examples of fine-quality luggage bear the name. The company also produced 'Cruzer' brand luggage, which name appears all too infrequently on vellum suitcases of the late 1920s and early 1930s.

Francis

Maker

Falmouth

Found impressed on early hide despatch-bag with strap.

Franklin, Iddins & Co. Ltd

22 Australian Avenue, London EC1

Leather case makers. No marked examples found to date.

James Fraser

280 Southampton Street, Camberwell, London SE5

Obscure, early 20th-century manufacturer of trunks and portmanteaux.

Frenchs Ltd

32 Moor Lane, EC3
57–63 East Road, N1
2, 4 & 6 Leonard Street, EC2

Given its various addresses, it is to be suspected that this company, advertising as trunk and portmanteau makers in the 1920s and 1930s at least, was reasonably prolific and successful. Since there do not appear to be any marked examples, it was probably an exclusively wholesale operation, similar to Penton's, supplying various retailers with luggage to order, and attaching retail labels to their goods as required.

H. Funston Benston

55 Clarence Street, Kingston

Trunk, bag and leather case maker. An oval, morocco label may be found on a variety of luggage, including 'Revelation' brand cases. On later examples the label may be of stamped, tan-coloured, 'mock-croc' paper, with the details printed in silver ink.

A. W. Gamage Ltd

118–128 Holborn, London EC1

Department store holding a Royal Warrant as Sports & Athletic Outfitter. Its sporting luggage – often square, suitcase-type (rather than soft-sided) leather tennis-racquet cases of well-patinated cowhide in a desirable shade, with good-quality, cast-brass locks – is marked 'Gamages'.

A. J. Garnett Ltd

50 & 52 Goswell Road, EC1
73 Long Lane, EC1

Probably a wholesale supplier to the retail trade, the company advertised as makers of attaché-cases, leather cases and suitcases.

R. & S. Garrard & Co. Ltd

1915 details:
24 Albemarle Street
17 Grafton Street, London W1
Manufactory:
1–3 Avery Row, Grosvenor Street, London W1
Est. in the Haymarket AD 1721

1915 Royal Warrant Holder, Crown Jewellers and Goldsmiths. (See Goldsmith's and Silversmith's Company.)

'Garrison'

('British Make Regd.')

Early-to-mid 20th-century brand of picnic sets of medium quality.

A. Garstin & Co. Ltd

158, 159 Aldersgate Street, London EC1

Whilst their catalogue stated unequivocally: 'We do not supply retail customers on our own account', Garstin's, a popular wholesale supplier of various types of luggage and leather goods to the retail trade during the 1920s and 1930s, advertised widely. Unlike many competitors, Garstin's promoted their own products in ways that made them easily recognisable, even where a separate retail trade label giving another supplier's name is affixed to the item. Anything bearing the 'Everwear – regd, TM' and logo was a product of this company. Additionally, the firm produced

Page from 1933 Garstin's catalogue

HAND LUGGAGE

Strong Morocco Grained Cloth

No. **Z8054.**—Horse-shoe shape Hat Bag, morocco grained cloth, leather hinge, stiffened sides, limp base and lid, lined with fine quality figured lining. Colours : black, brown, navy, olive.

14″	16″
8/6	**9/6**

No. **Z8107.**—Morocco cloth Overnight Case, en suite with Hat Box. Two nickelled slide locks, strong handle to match, lined with fine quality figured lining.

16″	18″	20″
8/6	**10/–**	**11/6**

HAT BAG. Special Line. Attractive Dice Grained Cloth

No. **Z8055.**—Stout dice grain cloth, best quality handle, with polished brass slide, two spring side fasteners, full size hinge, soft lid and base, covered metal frame, lined with fine quality striped art silk, drawn pocket at back.

14″	16″	18″
13/–	**15/–**	**17/–**

No. **Z8103.—Overnight** Case in dice grain cloth, as Hat Box No. 8055. Two lift-up spring fasteners at side, two polished brass slide locks on front, handle covered with same material as case, covered metal frame, lined with striped art silk lining, three pouch drawn pockets at back, choice finish.

16″	18″	20″
15/–	**17/–**	**19/–**

BENTWOOD HOOPS ON THREE-PLY FOUNDATION

Fitted with detachable wire cones

Covering of stout canvas, fitted with one lock, two clips, strong frame, leather handle, striped lining, four wire cones.

No.		Price, each
8523.—Brown painted canvas		
	Size 18 × 16½ × 16½ ins.	**27/–**
	,, 20 × 17½ × 17½ ins.	**29/–**
8524.—Green rotproof canvas		
	Size 18 × 16½ × 16½ ins.	**29/–**
	,, 20 × 17½ × 17½ ins.	**31/–**

distinctive 'Klip-it' Patented/ Registered locks and these are clearly stamped 'Garstin's Pat 25031/13' and/or 'Rd 543043'. Whilst not eminently collectable in its own right, a large quantity of excellent, medium-to-fine quality luggage survives in good condition, still suitable for practical use.

Ernest Gems

17 Queen's Gate Terrace, London SW7

Trunk and portmanteau maker in the early decades of the 20th century. No marked examples to date.

Gibaud & Higley Ltd

Wimbourne House, New North Road, London N1

Dressing-case makers in the 1920s and 1930s. Probably exclusively wholesale suppliers to the retail trade.

Charles Walter Gilbert

Crown Yard, 65 1/2 Stanhope Street, London NW1

Leather case makers of the 1930s.

Gildesgame Bros. Ltd

Importers

2 & 3 Roscoe Street, London EC1

Reptile-skin merchants of the 1920s/30s.

Arthur Gilmore

New York, USA

Superb trunks, similar in conception and execution to those produced by Louis Vuitton. Seems to have been most prolific in the first third of the century.

'Globe-trotter'

Founded 1897. Cases bearing the logo, reminiscent of an ivy leaf, and name of this famous brand are still produced in volume. Early examples are rarely available for purchase, as their long-term owners are loath to dispose of them. Extremely durable, practical, and increasingly collectable.

S. Goff & Co.

1923 address: 17, 18 & 22 King Street
28 Bedford Street, Covent Garden, London WC2

Advertising as the 'House for Economy – Colonial and Tropical Outfitters Civil and Military Tailors' the company lists the following examples in 1923:

'PRICES MOST MODERATE
QUALITY THE BEST
Cabin Trunks . . . from £1. 15. 0
Zinc-lined chests . . . from £4. 4. 0
Suit Cases . . . from £1. 2. 6'

By the 1930s the addresses had changed slightly, listed as 17 & 18 King Street, Covent Garden, WC2 and 25 Tottenham Court Road, W1. A variety of alternative addresses may also be found on applied labels: 215 Tottenham Court Road; 62 Seymour Street, Marble Arch; 302 High Holborn WC2. Examples are scarce.

Goldpfeil

Current details: Kaiserstraße 39–49
63065 Offenbach, Postfach 10 06 62, Germany
Tel: 00 49 (0)69 80500

Founded in 1856 by the pursemaker Ludwig Krumm, together with his five sons, to produce wallets, purses, bags and luggage. Soon established a name for excellence, its reputation for goods of superior quality and superb craftsmanship rapidly spreading, leading to orders for export to Russia and Poland. From 1872 the Krumms opened up the markets of the British Empire, starting with England. By the turn of the century 1,000 staff were employed, making the undertaking the world's largest in this industry.

In 1922, Heinrich Krumm, a grandson of Ludwig, left for New York with the aim of conquering the American market. Renamed Goldpfeil ('Golden Arrow') the company scored the bull's-eye at the Universal Expositions in Paris (1937) and Brussels (1958), being awarded Gold Medals for outstanding products at each event.

After setbacks caused by the war, Goldpfeil was again exporting to 35 countries by 1950. At present this truly international brand is represented in five countries by fifty Goldpfeil shops, taking particular pride in a commitment to tradition, craftsmanship and aesthetics.

Goldsmiths & Silversmiths Company Ltd

112 Regent Street, London W1

Listed in trades directories in the early decades of the 20th century as dressing-bag makers, along with the expected entries as jewellers etc., the company was later absorbed by Garrards. Few, but exceptional, examples of fitted dressing-bags and cases are still in existence, usually much cherished and therefore in superb condition. Almost every surviving example retains the original canvas foul-weather cover.

Charles George Goord & Sons

44–47 Colebrooke Row, Islington, London N1

Attaché, dressing and writing-case maker. It is interesting that for a period in the 1930s, the only other maker to emphasise its production of writing-cases in advertisements was a near neighbour, G.T. Dearberg & Sons Ltd, which suggests a strong element of competition between the two companies at that time.

Alfred Gough

28 Boulton Road, London W1

Trunk and portmanteau maker of the 1920s/30s.

Gough Kidston & Co.

43–45 Great Tower Street, EC3
1 Bermondsey Squar, London SE1

This company was the only one listed under 'Valonia Importers' in the *London Trade Directory 1934*. Valonia, occasionally spelt 'Vallonia', was the acorn cup of the Greek or Turkish oak, sometimes used in the production of tannin.

Goyard

233 Rue St Honore, Paris; Monte-Carlo; Biarritz

Founded in 1853, and still trading as a seventh-generation family concern. Produced a wide range of trunks and cases covered with their own distinctive fabric – brown canvas woven with a 'diamond' trellis pattern – and with chequerboard canvas similar to Louis Vuitton's 'Damier' fabric.

Graham

40 Gresham Street, London

Identified by the mark 'Graham – Makers' and the above address stamped directly onto the hide, this name has only been encountered to date marked on early-to-mid Victorian coaching pieces, such as a simple, lidded box (two pieces, not hinged, but designed so that the lid slid onto the base) held together with a strap – typical dimensions approximately 12 x 10 x 18 inches. It is possible that the company went on to produce more modern standard luggage, but given the absence of later marked items this is unlikely. Rare and interesting, if impractical for use as luggage in the present day.

Leckie Graham

95 Renfield Street, Glasgow

Maker of excellent, classic cowhide luggage. Cast-brass locks are often stamped 'secure lever – Made in England' and the applied trade label may state 'Bag and trunk makers'. Along with A. Boswell, Cleghorn, Forsyth, Irving Brothers, Reid & Todd, one of the prolific producers of fine leather luggage in Scotland who consistently marked their products.

Gray's Inn Trunk Stores Ltd

7, 9 & 11 Gray's Inn Road, London WC1

Although advertising in the early 1930s as trunk and portmanteau makers, given the central address it is likely that this company was a retail outlet that contracted-out orders to small producers in less commercially active areas.

H. Greaves

35 & 36 New Street, Birmingham

(Later H. Greaves Ltd.) Superb quality. (See illustration below of Gladstone bag, the lining stamped in gold leaf: 'H. Greaves est 1720 – by her majesty's Royal authority – dressing-case, bag and portmanteau makers, New street, Birmingham – prize medal for general excellence'.) A later example – a large, fairly standard, apparently run-of-the-mill suitcase – opens to reveal a honey-coloured pigskin lining, gold-tooled and stamped with a similar inscription. Expect heavy, cast-metal fittings. Locks may be marked 'Secure lever' and 'Greaves Ltd. Birmingham'. Not as widely renowned as such makers as John Pound, Finnigans or Drew & Sons, but consistently produced luggage of excellent and enduring quality.

Crocodile Gladstone bag lined with dark-green-morocco leather *c.*1880 by H. Greaves.

Herbert Gutteridge

171 Holland Road, London W14

Trunk and portmanteau maker advertising in the 1930s. Little known.

Guy & Son

Makers

Walsall

Finest heavy-duty quality, late 19th-century, hand-stitched luggage. Usually marked with the above information and possibly the words 'warranted hand sewn', quirkily stamped upside-down and frequently askew on each side of the lid.

G. W. S. & CO.

See G.W. Scott & Sons Ltd.

George W. Hall

The Derby Bag Stores

Corn Market and London Road, Derby

Trade label with this address and the words 'Trunk, bag and Portmanteau Manufacturer' occasionally found applied to the interior of early 20th-century suitcases etc.

J.B. Halley & Co. Ltd

Granville Works, Granville Square, London WC1

'Actual makers of golf-clubs, golf-bags and accessories'. This early 20th-century golf specialist produced leather golf-bags.

T. Hanford

An early Victorian manufacturer. As the trade label on one small, fragile, but nonetheless heavyweight, wood-lined, brass-bound, leather trunk with individually made brass lock proclaims:

'T. Handford Patentee
of the New Invented
Light Water Proof Travelling
Trunk

particularly adapted for Officers of the Army & Navy

At his Manufactory
No 6 Strand near Charing Cross
LONDON

Solid Leather Portmanteaus
Carpet, Coat & Enamell Bags
HAT CASES &C.'

Probably a rare survivor from the pre-railway era, designed particularly for travel by horse-drawn coach. Whilst compact, it was extremely heavy by later standards, and not designed to withstand the type of impersonal treatment that was likely to be meted out by railway porters. With the advent of widespread rail travel, the first drastic changes in the design of luggage took hold.

Hardy Bros. (Alnwick) Ltd

Mid 1930s address: 61 Pall Mall, SW1

World-renowned makers of fishing tackle – rods and reels in particular, Hardy Brothers additionally produced fishing-tackle boxes of outstanding beauty, strength and quality. Also highly collectable are fishing-fly wallets with this maker's mark – usually of cowhide, but see illustration on p.43 of rare example in crocodile.

Harman & Son

(late of 87 New Bond Street)
65 Jermyn Street, SW (one door from St James Street)

The above found ornately gold-blocked with a pictorial logo onto the green-morocco interior lid of a rare, early 20th-century cube hatbox with front handle. (Patent No 286801, Replex. Regd.)

Alfred Harrison

77 Old Street, London EC1

Leather case makers. No marked examples found to date.

Harrods Ltd

1930s listing:
87–135 Brompton Road, SW1
62 & 64 Brompton Road, SW3
Hans Road & Basil Street, SW3
Hans Crescent, SW1
Draycott Avenue, SW3
Sloane Avenue, SW3
Telegram Address:
'Everything, Harrods, London'
Current listing:
Harrods, Knightsbridge, SW1
Tel: +44 (0) 171 730 1234

Royal Warrant Holders

Founded in 1849. Amongst 'everything', this world-renowned store has sold every type of luggage imaginable down the years. Particularly sought after are those items that were formerly produced by the Harrods Luggage Manufactory, and which frequently bore the name, gold-blocked directly onto the fabric of the case.

Harrods still has an extensive Luggage Department, stocked with the finest variety of top quality luggage available, made by many of today's most highly regarded manufacturers. It is comforting, should one be unable to find the perfect antique item, that much of the luggage stocked by Harrods is currently still made in classic style to traditional standards.

Hartmann Trunk Company

USA

World famous for its substantial luggage, often covered with mottled, striped canvas in muted autumnal shades, or grey and red – though many different types exist. Most notable for impressive wardrobe-trunks.

W. Harvey

Saddler

Stockport

Above mark impressed directly onto base of cases produced by this maker, examples of which are rare.

George Edward Haslam

32 Northampton Square, London EC1

Attaché-case maker in the late 1930s.

Haydon

Maker

Bristol

Rarely found West Country maker. Appear to have shared local traditions of excellence with such companies as Insall and Bracher. The use of top-quality bridle hide and heavy-duty brass fittings echoes the products of these other Bristol-based firms, as does the practice of impressing the above details on each lateral side of the lid. Haydon stamped this plainly, with a decorative flower motif on either side of the name.

Hedges

Maker

High Street, Leicester

This name has been found stamped under the top handle and onto the end of each 'fish-tail' handle attached to the separate lid of an early and rare Victorian hide travelling box made in two pieces, with elaborate, hand-tooled decoration.

Hepburn & Cocks

11 Sheffield Street, Lincolns Inn Fields, London WC2

Manufacturers of despatch-boxes, established 1790.

'Hercules Make'

(Guaranteed stiffened, upon compressed fibre)

Stamped-out brass retail label found on wooden-hooped, canvas-covered trunks, early-to-mid 20th century.

Oskar Hermann

'Spezialgeschäft fur Koffer u Taschen'

Bismarkplatz, Dresden, Germany

Wooden-hooped, canvas trunks with exquisitely detailed painted decoration, generously embellished with brass fittings and studs, are amongst the rarely found products of this late 19th-century specialist in trunks and bags.

Hermès

Current details:
24 Rue de Fauborg St Honore, Paris
Hermès (GB) Ltd Head Office:
176 Sloane Street, London W1
Shops:
155 New Bond Street, London W1
3 Royal Exchange, London EC3
179 Sloane Street, London SW3

Founded in 1837, Hermès has long been world famous for the quality of workmanship of all products bearing the name. Especially of interest to collectors of antique luggage are superb dressing-cases, often with an integral, smaller case containing the fittings, which rests open within the larger case like a tray, with contents displayed ready for use, and which can be lifted out, closed and carried separately. Another highly covetable Hermès creation is the distinctive 'Kelly' handbag.

M. & A. Hess Ltd

38 Tyer's Gateway, Market Street, London SE1

Trunk, portmanteau, attaché, dressing-case and suitcase makers in the 1930s.

The Heston Aircraft Co. Ltd

Trading Estate, Slough, Bucks

A post-war advertisement for this company offered 'light, lively luggage for the traveller of taste'. The description states that 'Astral' brand luggage is made from a special aeroplane alloy:

> 'Which can't sag, rust or wear out. Astral is the AIRSTREAMED luggage; just right for airline travel – light as air to carry – streamlined in design.
>
> GAY HOLIDAY COLOURS
>
> Both for charm and convenience Astral present matching sets of suitcases in 9 glorious colours. They are lined with velvety 'Spraytex' which won't trap dirt or tear, can be washed… Astral sets are sold wherever travellers of taste buy luggage, in these 9 SMART COLOURS, linings of the same or contrasting tones:
>
> MIDNIGHT BLACK
> AUTUMN BROWN
> OXFORD BLUE
> NILE GREEN
> CAMBRIDGE BLUE
> DEEP IVORY
> GRENADIER RED
> LAUREL GREEN
> MAROON.'

Leslie Heynemann & Co.

52 & 54 Weston Street, Bermondsey, London SE1

Reptile-skin merchant in the 1930s.

Hill & Millard

103 Jermyn Street, London SW1
7 Duncannon Street, Trafalgar Square, London

Famous 19th-century military outfitters. Advertised as trunk and portmanteau makers. Their finely engraved, paper trade label is a sought-after addition lending added appeal. The variety of styles and quality of the admittedly fairly rare pieces does, however, tend to lead to the conclusion that the company perhaps bought in a range of stock from wholesalers and applied their distinctive label – since unlike most other highly regarded names, there do not appear to be any particular trademark characteristics. The Duncannon Street address is likely to be found directly impressed onto the upper-front, exterior-base section of small leather articles.

Tom Hill (Sloane Square) Ltd

44 Sloane Square SW1; 3 & 5 Holbein Place, SW1 London

Very rare and highly sought-after maker of finest quality baggage, on a par with Lansdowne Luggage, or Peal & Co. Usually linen-lined, always heavy, cast-brass fittings. Exquisite quality and construction.

Hindmarch Bros. Ltd

52, 54 & 56 Ormside Street, London SE15

Trunk, portmanteau, attaché and leather case makers in the first third of the 20th century.

J. Hodges

Maker

Above name occasionally found stamped on good-quality, cast-brass locks. It is not clear whether the term 'maker' refers to the fittings alone, or to the entire case. Occasionally an applied trade label, usually that of a department store such as A. & N.C.S. Ltd may also be found, which suggests that J. Hodges might have been a wholesale supplier to the retail trade.

Holland & Holland Ltd

1930s addresses: 98 New Bond Street W1
Factory: 906 Harrow Road W10
Shooting grounds: Northwood, Middlesex
'By appointment to: His Majesty the King; HM The King of Italy; HM The King of Spain etc. etc.'

These gunmakers are still in existence, their factory listed at the same address (in the re-defined postal area NW10) along with 'Duck's Hl Rd, Northwood (Shooting School)'. The retail premises are now 31 Bruton Street, London W1. Supplied lovely oak-lined cartridge-cases and gun-boxes with heavy, cast-brass fittings that included reinforcement corners. Also currently carry fine vintage luggage along with their own exclusive designs in flax and leather.

W. Honegger

Lausanne
(En face de l'ecole Vinel)

Late 19th-century Swiss trunk and case maker, influenced by Louis Vuitton. Items marked on exterior with stamped-brass trade label.

Houghton & Gunn

161 New Bond Street, London

Founded 1822. High-class leather-goods manufacturers and stationers, bought out by Asprey & Co. in 1906.

W. Houghton

54 & 56 Southampton Row, London WC

The above, and the word 'maker' usually found stamped black above handle on leather suitcases typical of the early 20th century.

Hyde, William & Co.

122 Euston Road, London NW1

Early 20th-century trunk makers.

'Innovation'

New York (242 5th Avenue), Chicago – Paris – London

Distinctive ladies' wardrobe-trunks, shaped with curvaceous bulges, usually brown or black in colour and opening vertically (see illustration on p.20). Marked with a pennant-shaped, exterior-applied, brass trade label giving above details plus: 'pat'd May 17 1898 – Jan 30 1900'.

Examples of a later wardrobe-trunk of the more standard style, which splits in half horizontally, bear a similar tag with the words 'patented March 5 1912'. This latter tag was found accompanying a label stating 'Made by Innovation Trunk Co for Gimbel Brothers New York', the leather carrying handles impressed 'Gimbel Brothers'. Other types of luggage can also be found bearing this name.

W. Insall & Sons

19 & 20 St Augustine's Parade, Bristol

Highly regarded and collectable West Country maker. Often decorated with exquisitely detailed hand-tooling, some very fine portmanteaux and cases in excellent condition survive and are highly collectable. The maker's name is invariably found stamped on each lateral side of the lid. (See Brachers.)

Irving Brothers

99 Princes Street, Edinburgh

Victorian supplier of luggage. Items are rare and frequently lined with blue-and-white striped ticking, edged with scalloped red-morocco leather. The trade label includes the words: 'Portmanteau and Brush mfrs To Her Majesty – military and marriage outfits'.

William Ives

8 Guildhall Street, Lincoln

Early-to-mid 20th-century leather, and canvas-and-leather luggage. A silver-blocked, rectangular, green-morocco label to the interior of the article is likely. The label abbreviates the name to 'Wm' and includes the information that the range available from the outlet embraced travelling bags and trunks and that William Ives was a sports outfitter.

W.E. Jackson

The Bag Stores

Nottingham

'Makers of Trunks, sample cases &c'. Late-Victorian and Edwardian cases bearing the large, black, oval, gold-blocked trade label of W.E. Jackson will usually also have the name impressed into the hide on each outer-lateral side of the lid. By the 1920s, the label may occasionally be found on cases by other manufacturers, such as 'Antler' brand items with Legge locks. These later trade labels, affixed to items not actually produced by W.E. Jackson but retailed at The Bag Stores, tend to be rectangular in shape, and silver embossed, still on black-morocco leather. Also retailed a range of picnic sets by other makers.

Finely tooled portmanteau by Insall *c.*1870.

Albert Jacobson & Co. Ltd

37 Strand, London WC2

1920s/1930s trunk makers.

Isidore Jacobson

4 Westbourne Grove, London W2

1920s/1930s trunk makers.

G.H. James & Co. Ltd

48 Old Bailey, London EC4

Early 20th-century dressing-bag maker.

Richard Attenborough Jay & Co. Ltd

142 & 144 Oxford Street, London W1

1930s dressing-bag makers.

Jenner & Knewstub Ltd

33 St James Street; 66 Jermyn Street, London

Exquisite despatch-boxes, wood-lined with morocco-leather covering. Maker's name gold-blocked, usually on the interior-front rim.

Jones Brothers (Holloway) Ltd

Trunk Department

A rectangular, 2 x 3 inch, gold-blocked morocco label, applied to the interior lid identifies a variety of luggage supplied by this well-known departmental store.

B. Joseph & Son

Travel & Sports Depot

Sunderland

Products purchased in the early years of the 20th century may be identified by the oval, paper label to the interior of the lid, silver on blue.

Gustave Keller

Rue Joubert, Paris, France

Lovely dressing-cases, finely crafted, often of delicate morocco in dark colours with vermeil-topped crystal bottles. Keller also produced a wide range of cased items, such as silver tea-sets, travelling kettles, picnic sets etc. Highly sought-after and collected, Keller is perhaps more recognised in Europe than in the UK and elsewhere.

Percival H. Kidson

28 James Street, Harrogate

A tiny, dark-red label, approximately 1/2 x 3/4 inch, printed gold, its shape reminiscent of a butterfly, states 'Leather and fancy goods'. Rare and usually found on small items such as writing-cases.

W.A. Kiernan

133 Regent Street, London

Though an obscure name, the rare products that do exist tend to be charming and of excellent quality. Usually pale honey-coloured cow-hide with heavy, cast-brass locks, lined with tan-coloured, textured morocco-leather lining. The brass locks may be stamped 'Double lever lock' above the slide, and 'KIERNAN LONDON' beneath.

Edward King

148 Farringdon Road, London EC1

Leather case makers, mid 1930s.

Henry George King

Leather

39 Chapel Road, West Norwood, London SE27

Advertised as a bag maker in the 1930s.

Alfred Ernest Kingsman

23 Brookfield Road, Hackney, London E9

Suitcase maker of the 1920s/1930s.

R. & A. Kohnstamm Ltd

10 & 11 Chiswell Street, E1
Tannery: Foxberry Road, Brockley, SE4 London

Reptile-skin merchants advertising aggressively in the 1920s/1930s.

F. Lansdowne

55 Jermyn Street, London SW
23 Piccadilly Arcade, London SW1

Collectable and fairly rare. Advertised as makers of suit and dressing-cases, dressing-bags, trunks and portmant-eaux. Items are scarce, usually standard linen-lined, light honey-coloured cowhide. Some examples are embossed 'F. Lansdowne' with the address, but many later examples are impressed:

'LANSDOWNE
LUGGAGE
LONDON'

Fittings are always very high quality brass or gold-covered brass, always cast (not stamped-out) metal. A few examples are impressed 'Lansdowne's "Inviolable" Lock' but usually the locks are also stamped with either of the two alternatives given above for the leather body of the case – normally whatever is impressed on the hide of the case matches the inscription stamped onto the brass locks. 'Lansdowne of Jermyn Street' has also been observed, on one example gold-blocked on the front rim under the lid.

The Lane Trading Co. Ltd

10 Idol Lane, London EC3

Reptile-skin merchant of the 1930s. It is interesting to note the proliferation of such brokers and wholesalers in the early years of the century, and their later demise.

S. Last

15a Grafton Street, London

Leather trunk makers. Lease of their property acquired by Asprey's in 1902. On later examples label may state '165 New Bond Street Corner of Grafton Street'.

Lavino (London) Ltd

48 Fenchurch Street, London EC3

Reptile-skin merchant, late 1920s/early 1930s.

Arthur J. Lawrance

136 Kensington High Street, London W8

Trunk and portmanteau makers, early- to-mid 20th century.

Lawrence & Sons

444 Strand, London WC2

Likely to have been a retail outlet – their trade label tends to appear applied to cowhide 'Revelation' brand expanding suitcases, along with various other types of leather suitcases.

Gavin Lawson

61 Fore Street, London EC2

Reptile-skin merchant, mid 1930s.

John Leader Ltd

18 Eldon Street, EC2
22 The Arcade, Liverpool Street, EC2

14, Terminus Place, Victoria, SW1
166 Strand, WC2

The shrewd placement of the latter three addresses near to Liverpool Street, Victoria and Charing Cross railway stations would suggest that this trunk and portmanteau maker was a thriving retail operation.

Leckie Graham

See under Graham

John Leckie & Co. Ltd

Imperial Buildings, 56 Kingsway, WC2
Factory: Walsall

Suitcase makers of good-quality luggage. The factory was based in Walsall, traditionally a centre for the manufacture of luggage.

G.H. Lee & Co.

Liverpool

Quite rare, dark, well-patinated leather cases. Appear to have specialised in very large suitcases with both side and front handles plus surrounding retaining straps.

Legge

The above, or 'Legge locks' along with a circular, three-legged logo, appears on suitcase locks produced by this company.

L.A. Leins & Sons

59 Shoe Lane, London EC4

Advertised as dressing-case makers in the 1930s.

N. Lessof Ltd

356 Kingsland Road, London E8

Suitcase makers in the 1930s.

W. Leuchars

38 & 39 Piccadilly London
2 Rue de la Paix, Paris

Established in Piccadilly in 1794 as a perruquier. When wigs went out of fashion, they became gold- and silversmiths, supplying dressing-cases, notably to Queen Adelaide, consort to William IV. Leuchars was a venerable maker in Victorian times (see illustration on p.114 of large, pigskin-lined leather trunk, and of blue-morocco-cased tea-set on p.62). It is recorded that Mr Leuchars 'would never demean himself by advertising his wares'. In 1884, Leuchars absorbed Louis Dee, a wholesale jewellers founded in 1830 by Thomas William Dee, and in 1888 W. Leuchars in London was taken over by Asprey's. Asprey's had also bought the Sherwood Street factory of Louis Dee and added the production of fine leather goods to its work as a silversmith. The Paris branch of Leuchars was taken over in 1888 by its former manager, Monsieur Geoffroy, who acted as the Paris agent for Asprey until 1902.

Lincoln Bennett

London W

Many square, leather, gentlemen's hatboxes from the early years of the 20th century contain a gold-blocked label indicating that they were supplied by this successful and prolific hatmaker. The applied label generally matches the colour of the lining.

Louis Vuitton

See under Vuitton

Henry James Lowe

58 & 59 Chiswell Street, London EC1

Advertised as leather case makers in the 1930s.

McBrine Baggage

A shield-shaped, paper label depicts the globe, circumnavigated by various trunks, bags and cases, reminiscent of Saturn's ring. Mid 20th century.

F. McMillan & Co.

51, 53 & 55 Bermondsey Street, London SE1

Advertised as attaché-case makers in the mid 1930s.

T. Mabane & Sons

Leeds

Maker's name found impressed on medium-grade, early-to-mid 20th-century cowhide suitcases.

Macy's

New York, USA

The retail label 'R.H. Macy & Co. New York' of this famous store may be found on a wide selection of luggage by various makers. Many wardrobe-trunks.

Moritz Mädler

German or Austrian maker, especially of fine-quality crocodile luggage. Late 19th century. Sought-after brand.

F.J. Mallia Ltd

25 & 26 Lime Street, London EC3

Reptile-skin merchants trading in the first third of the 20th century.

W. Mansfield & Sons

90 Grafton Street, Dublin

Royal Warrant Holders in the second decade of the 20th century as dressing-bag makers.

Large, pigskin-lined, cowhide trunk by Leuchars, *c.*1870.

Maple & Co. Ltd

Tottenham Court Road, London W1

Famed for furniture. In the 1920s and 1930s a few exquisite and elaborate fitted wooden picnic sets, formed as a box with brass catches and side carrying handles, with fold-out flaps and legs were produced by and bear the trade label of Maple & Co.

Mappin & Webb Ltd

Jewellers, Goldsmiths, Silversmiths

Current Head Office:
413 Oxford Street, London W1
Tel: +44 (0) 171 409 3377

Founded in 1774 by Jonathan Mappin, who rapidly became known as one of the foremost silversmiths in England. Now primarily known as jewellers and silversmiths with many branches in the United Kingdom and in Paris, Dusseldorf, Johannesburg and Tokyo. In the 19th and early 20th centuries, many exquisite dressing-bags and cases with fittings of gold or silver were produced by Mappin & Webb, and are often still discovered in excellent condition, protected by a canvas cover. Mappin & Webb bought the Alexander Clark Company in the mid 1960s. The following list of addresses, taken from directories, may assist in the dating of items of luggage:

JOSEPH MAPPIN

1846–49	*15 Fore Street*

JOSEPH MAPPIN & BROTHERS

1850	*15 Fore Street*
1851–55	*37 Moorgate Street*
1856–57	*67/8 King William Street*

MAPPIN BROTHERS

1858–61	*67/8 King William Street*
1862–69	*222 Regent Street* *67/8 King William Street* *400 Euston Road*
1870–71	*220/2 Regent Street* *67/8 King William Street* *400 Euston Road*
1872	*220/2 Regent Street* *67/8 King William Street*
1873–79	*220 Regent Street* *7/8 King William Street*
1880–85	*220/2 Regent Street* *67/8 King William Street*
1886–88	*220 Regent Street* *67/8 King William Street*
1889–90	*220 Regent Street* *35 St Paul's Churchyard*
1891–92	*220 Regent Street* *66 Cheapside* *(late King William Street)*
1893–1903	*220 Regent Street* *66 Cheapside* *(late King William Street)*

Incorporated with Mappin & Webb, Ltd

1904	*220 Regent Street*

MAPPIN & CO.

1860–63	*77/8 Oxford Street*

MAPPIN, WEBB & CO.

1864–67	*77/8 Oxford Street* *71/2 Cornhill*
1870–71	*76/7/8 Oxford Street* *71/2 Cornhill*
1872–73	*76/7/8 Oxford Street* *20/1 Poultry, Mansion House Buildings*
1874–75	*76/7/8 Oxford Street* *Mansion House Buildings, 2 Queen Victoria Street*
1876–77	*76/7/8 Oxford Street* *Mansion House Buildings, 2 Queen Victoria, Poultry*
1878–79	*76/7/8 Oxford Street* *2 Queen Victoria Street*
1880	*76/7/8 Oxford Street* *Mansion House Buildings, 2 Queen Victoria Street*
1881	*76/7/8 Oxford Street* *2 Queen Victoria Street*
1882–85	*158–162 Oxford Street* *Mansion House Buildings*
1886–87	*158–162 Oxford Street* *18/22 Poultry (Mansion House Buildings)*
1888–89	*158–162 Oxford Street* *34 King Street* *2/4 Queen Victoria Street* *18/22 Poultry*
1890–96	*158–162 Oxford Street* *2/4 Queen Victoria Street* *18/22 Poultry*
1897–98	*158–162 Oxford Street* *2/4 Queen Victoria Street*

MAPPIN & WEBB LTD

1899–1903	*158–162 Oxford Street* *2/4 Queen Victoria Street (Mappin Brothers Incorporated)*
1904–15	*158–162 Oxford Street* *220 Regent Street* *2/4 Queen Victoria Street*

Marris's Ltd

101 Hatton Garden, London EC1

Prolific producers of 'Sirram' Brand picnic sets. ('Sirram' is a reverse spelling of the company name.) Most were of medium quality.

Marshall & Snelgrove

334–354 Oxford Street, London W1

Cases and trunks with this applied trade label are occasionally found. Marshall & Snelgrove had other branches, including one in Birmingham. The label is usually a long, thin rectangle, rounded at the ends, made of ivorene, approximately ¼ x 2 inches.

A. Marshall

Newcastle Place, Edgware Road, London W2

Recorded as trunk & portmanteau makers, mid 1930s.

Marsh's Trunk Makers

10 King Street, Manchester

Trade label found mainly on attaché-cases. Medium-to-high quality.

A.L. Martin & Co.

18 Market Street, Bermondsey, London SE1

Reptile-skin merchants trading in the first third of the 20th century.

J. Mason & Co. Ltd

263 Finchley Road, London NW3

Recorded as trunk and portmanteau manufacturers in the mid 1930s.

William Henry Matthews & Sons

46 & 47 Beech Street, Barbican, EC1

Makers of school satchels in the 1920s and 1930s.

G. Maude

Maker

Leeds

Maker's name found on finely hand-tooled leather attaché-case with squared patent locks, *c.*1925.

Mawson Swan & Morgan

Newcastle on Tyne

Oval maker's stamp, approximately 1 x ¾ inch, directly impressed onto front interior, beneath the lid. Usually linen-lined. Early 20th century. Advertised as silversmiths and dressing-bag makers.

'Migrator'

Patent No 310105

'A wardrobe in a hat box'.

Brand name owned and produced by A. Garstin & Co. Ltd, 1930s. (See illustration on p.84.)

Alfred Miller

181 New North Road, London N1

Advertised as trunk and portmanteau makers 1920s/1930s.

J.C. Moody

18 Tontine Street, Folkstone E Phone, 921, Kent, England

Rare but not particularly sought-after as cases are of medium-to-good quality only. Gold-blocked, oval, black, applied trade label states: 'Trunk Suit Case and Leather Warehouse' *c.*1920.

H. Moores

Saddler

Cardiff

Standard-size, high-quality, saddler-made suitcases. One example had

MAPPIN & WEBB

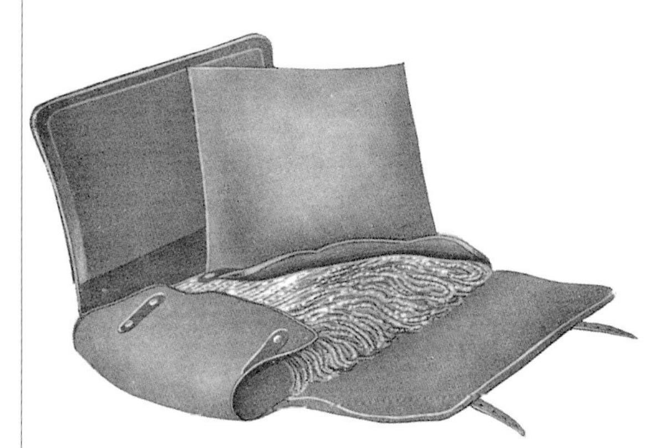

G 2270.—Levant Morocco Leather Folding Case, containing Cushion and Wool Travelling Rug .. **£8 15 0**

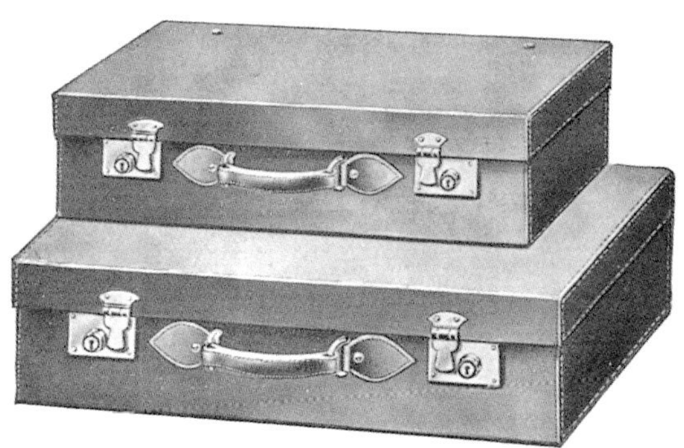

Gentlemen's London Made Hand-stitched Suit Cases.

	L 1930 Fine Hide Lined Drill.	L 1935 Best Hide Lined Leather.
20 in. × 14 in. × 7 in. ..	**£2 17 6**	**£6 15 0**
22 in. × 14½ in. × 7½ in. ..	**3 3 0**	**7 10 0**
24 in. × 15 in. × 8 in. ..	**3 10 0**	**8 5 0**
26 in. × 16 in. × 8 in. ..	**3 15 0**	**9 0 0**
30 in. × 18 in. × 8½ in. ..	**4 10 0**	**10 10 0**

Fitted with lever locks.

L 1988.—Lady's empty Dressing Cases in Fine Morocco. Lined Art Silk. Loose drawn pockets for own fittings.

18 in. × 13 in. × 6 in.	**£3 17 6**	Waterproof Cover	**£1 1 0**
20 in. × 14 in. × 7 in.	**4 7 6**	,, ,,	**1 2 6**
22 in. × 14½ in. × 7 in.	**4 17 6**	,, ,,	**1 5 0**
24 in. × 15 in. × 7 in.	**5 7 6**	,, ,,	**1 7 6**

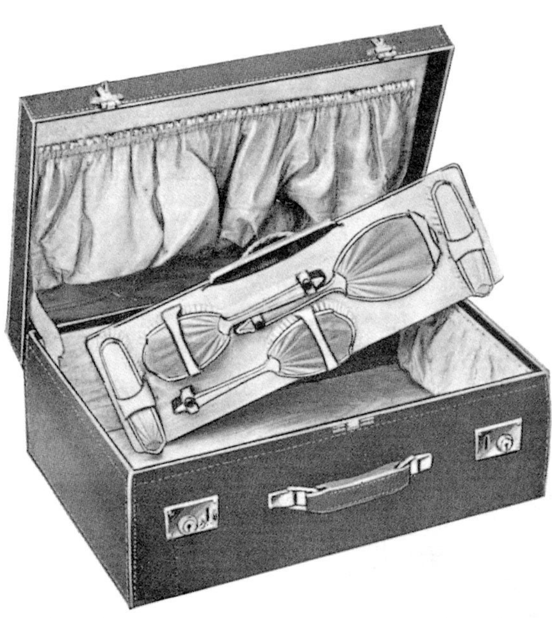

L 1970.—Lady's 22 in. Morocco Leather Travelling Case. Lined Art Silk and fitted with fine Enamel (newest colours) and Sterling Silver 6-piece Dressing Table Service on removable panel at back. The case has loose drawn pockets at the ends and in the lid for extra fittings **£22 5 0**
Case, as above, but without Dressing Table Service **8 15 0**
Waterproof Cover, **£1 5 0** extra.

'Wilkinson Patent' locks, and was lined with tan-coloured, mock-crocodile-stamped leathercloth, suggesting that this company was trading in the 1920s and 1930s.

S. Mordan & Co.

London

Nineteenth- and early 20th-century Despatch-Case Makers to H.M. Government. Perhaps best known as a stationer, most particularly for pencils.

H. Moss Jn.

159 & 161 Ball's Pond Road, London N1

Recorded as trunk and portmanteau makers *c.*1930.

Moynat

Shops: 1 Avenue de L'Opera
5 Place du Theatre-Francais
Works: 15 Rue Coysevox, Paris, France

Early 20th-century French brand name. Locks are routinely stamped 'MOYNAT'. Additionally, each lateral lid section may bear an embossed-brass, applied label in the shape of an inverted equilateral triangle, subdivided into four smaller triangles. The central one gives details of the company name and address, surrounded by a depiction of a plane flying past a ship, a train, and a motor car. A similar paper label, printed black on white, may be pasted on the interior lining.

Opposite
Typical cases from the 1934 Mappin & Webb catalogue.

J. Mullins Ltd

Fibre

78 Southwark Street, London SE1

Suitcase maker advertising in the late 1920s/early 1930s. The word 'Fibre' indicates that this company possibly created some of the many unmarked examples of vulcanised-fibre suitcases that are still used and enjoyed today.

Murray & Son

16 Broad Street, London W1

Recorded as dressing-case makers, *c.*1930. No marked examples encountered to date – surprisingly given the location, which would suggest a reasonably high-profile retail outlet.

Murrell & Son

12 Avery Row, Grosvenor Street, London W1

Trunk and portmanteau makers advertising in the mid 1930s. The location of the premises would indicate a high-profile retail outlet but no marked examples discovered.

Henry A. Murton Ltd

6–12 Grainger Street, Newcastle-on-Tyne

Above appears on an embossed, gold-on-cream-paper label applied to the interior lid of medium-quality, early-to-mid 20th-century leather suitcase. Label is of rectangular shape with canted corners.

National Trunk Co.

55 Praed Street, London W2

Advertised as trunk and portmanteau makers in the mid 1930s. Located in close proximity to Paddington Railway Station. Few marked examples. The company probably contracted-out commissions to small leather-goods businesses – it is most probable that this was primarily a retail outlet.

J.T. Needs

See Bramah & Co.

Newbury & Johnson Ltd

London SW9

Makers of shaped motoring-cases.

J. Nigst & Sohn

Wien, Austria

Usually marked on heavy, cast-brass locks. Superb-quality cowhide cases with distinctively stitched handle on brass mounts. (See illustration on p.69, second from top).

Norfolk Hide

Often stamped 'Ashtona STANDARD' or possibly marked 'Norfolk Hide' along with the patent number '134403, WARRANTED SOLID LEATHER'. A truly unique type of leather luggage, rare and highly collectable. Collar-boxes, large cases and hatboxes are extremely scarce. It is more likely that small A4-sized cases might still be found for sale. Interestingly, Norfolk hide takes its name from the fact that it was produced in Norfolk Road, Sheffield – and not, as one might assume, in Norfolk. Many sample-cases were produced in Norfolk Hide, notably for Terry's of York, chocolate manufacturers, apt given that Norfolk Hide is usually of similar hue to the products these items were designed to transport.

John Norris

158 Malden Road, London NW5

Obscure trunk and portmanteau maker trading in the 1930s.

S. Noton Ltd

Endurance Works, Blackhorse Lane, Walthamstow, London E17

Advertised aggressively in the 1930s as suitcase and attaché-case makers. Products are of medium quality and usually identified by the impressed mark 'Noton'. (See Attaché Case Manufacturing Co. Ltd and Suit Case Manufacturing Co. Ltd.)

Old England

Current details:
12 Boulevard des Capucines, 75009 Paris
Tel: + 33 1 47 42 81 99

Very rare examples of luggage from the early years of the 20th century bearing the maker's name 'Old England' survive, one a large, cube-shaped, pale-green canvas hatbox, leather trimmed. The store now stocks a charming selection of vintage luggage.

'Orient Make'

Brand name of high-quality, vulcanised-fibre trunks and suitcases, an attractive, lightweight and low-maintenance alternative to leather luggage that still embodies the charm of a bygone era. The parent company that produced this plentiful and durable brand of luggage and held copyright of the name and logo was The Bag Stores, Northampton. Most 'Orient Make' suitcases are lined with unbleached linen, and the motif of a rising sun beneath the name is jacquard-woven into the fabric on the right-hand section of the lid. The handle is usually constructed on a metal base. Locks are of cast metal. Many 'Orient Make' suitcases are still in frequent, heavy-duty use some 70 years after production. Items in exceptionally original condition may still bear an applied transfer, predominantly pale blue, black and gold, on each lateral side of the lid section, and a similar paper label to the interior mid-back base section either pale blue and dark blue or red and black. In addition, a separate applied label indicating the specific retail outlet to have supplied the piece was frequently considered appropriate by the original vendor.

Osh Kosh Trunk Co

USA

American manufacturer famed particularly for metal, brass-bound wardrobe trunks of the 1920s and 1930s. A popular retailer whose trade label may additionally be found attached is Marshall Field & Co. of Chicago.

'Ossilite'

Brand name found on lightweight luggage, early-to-mid 20th century.

'Pakawa'

Brand name found most often on brief-cases and bowls-cases, early-to-mid 20th century. Pakawa registered a spring-loaded handle, (similar in concept to that patented by Asprey) under patent No. 292680. This lay flat whilst not in use, the fixings concealed beneath a square of metal at each extremity. Not every brief-case possesses it but the products are almost always attractive, if in the main slightly more modern than many other brands discussed here.

Parker, Wakeling & Co. Ltd (Wardrobe)

13 Playhouse Yard, Golden Lane, London EC1

Trunk and portmanteau makers of the 1930s, apparently specialising in wardrobe-trunks.

T. Parry & Son

High Street, Wrexham

Blue paper label applied to interior lid states 'For saddlery, harness, trunks, bags & leather goods'. Early-to-mid 20th century.

Parsons, Jn. & Sons

157 Edgware Road W2; 185 High Road Kilburn NW6

Trunk and portmanteau dealers in the late 1920s and 1930s. Probably an outlet for the many obscure manufacturers who did not mark their products.

Patterson & Stone

Manufacturers of golf bags

Crichton Works, Frederick Street, Walsall

Manufacturer of golf-bags based in the leather-working centre of Walsall.

'Paxmor'

A rectangular, orange-brown, paper label applied to the interior linen-lined lid section, with a swag printed darker brown announcing:

'Expanding
The Paxmor
Suitcase Pat No 237458'

Similar in concept to 'Revelation' brand expanding luggage. The fixings are stamped 'THE PAXMOR Pat No 240050/25', and locks may be stamped 'BRITISH MADE 238734'.

'Paxwell'

Travel Goods (Registered)

'Guaranteed the West's best'

Made in Western Australia

Mid 20th-century brand. Large, circular, brown-on-cream paper label, approx 3 inches in diameter.

Peal & Co.

487 Oxford Street N; 47 North Row, Grosvenor Square, W1

Advertising as retail boot and shoe makers, and Royal Warrant Holders as bootmakers in the second decade of the 20th century. A very few rare and delightful examples of well-patinated, top-quality, hand-stitched cowhide luggage produced by Peal & Co. survive. Both boots and luggage are desirable and collectable. Luggage is invariably stamped 'Peal & Co.' across the front exterior rim of the lid. Heavy-duty, cast-brass locks and fittings.

Pearson & Pearson

Trunk & Portmanteau Makers

12 Angel Row, Nottingham

Classic cowhide, fitted cases, produced in the early years of the 20th century.

E.J. Pearson & Sons Ltd

Wholesale

Richbell Works, Emerald Street, Theobald's Road, London WC1

Established in Clerkenwell in 1804. Advertised as actual manufacturers of a huge variety of leather goods, and listed as leather-goods manufacturers, bag, leather case and suitcase makers ('Wetresista' patents being the proud claim for the latter). The range embraced 'solid leather & blocked cases for traveller's samples, scientific instruments, surveyor's rules, flask cases, clock, watch & cigar cases, patentees of "Hinfol" folding watch cases; attache, suit and music cases, letter, note & document cases, despatch-cases; Top-hole treasury & season cases; leather instrument & manicure rolls, cigarette tube cases, tobacco pouches'. A popular brand of luggage produced by this company was 'Victor' luggage, and many items of interest to collectors in other fields are contained in receptacles produced by this prolific maker.

John Peck & Son

Nelson Square, Blackfriars, London

Rare, early 20th-century despatch-boxes and brief-cases gold-blocked with name and 'Manufacturers to HM Stat. Office'.

Peel & Co.

62, 64, 66 & 68 Valetta Road, East Acton, W3

Bootmakers. Luggage very rare. Not to be confused with Peal & Co.

'Pendragon'

Brand name frequently found on medium-quality brief-cases and small, suitcase-style document-cases, usually gold-blocked with the words 'Made in England, TOP GRAIN COWHIDE, A Pendragon Product', occasionally on an oval, ivorene trade label. This is possibly a post-war brand name; however, the existence of classic 1930s-style brief-cases and a few attaché-cases with typically 1930s alloy locks suggests that the company was in existence earlier. Parent company not established to date.

Items by Peal & Co. (bootmakers), the large case probably designed to carry boots, the smaller one a cartridge-case made for The Lord Ampthill.

LEATHER KIT BAGS

SN 1099

SN 1099 Leather Kit Bag, covered steel frame, brass fittings, registered lock.
Made in sizes 16 18 20 22 inches
@ 26/8 30/– 33/4 36/8 each

SN 1101 Leather Kit Bag, covered steel frame, brass fittings, registered lock,
Made in sizes 16 18 20 22 inches
@ 20/8 23/3 25/10 28/5 each

BRIEF BAG

SN 1168 Tan Hide, covered frame, brass fittings. Made in sizes 12 14 16 18 20 inches
@ 16/– 18/8 21/4 24/– 26/8 each

COLLAR BOXES

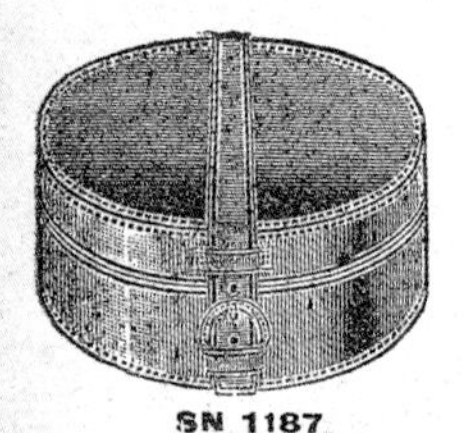

SN 1187

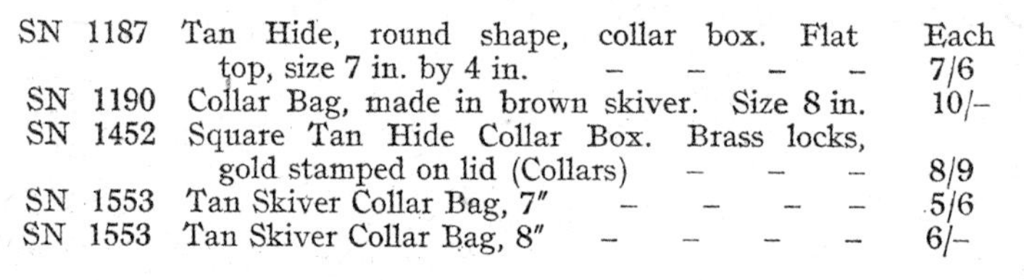

		Each
SN 1187	Tan Hide, round shape, collar box. Flat top, size 7 in. by 4 in. – – – –	7/6
SN 1190	Collar Bag, made in brown skiver. Size 8 in.	10/–
SN 1452	Square Tan Hide Collar Box. Brass locks, gold stamped on lid (Collars) – – –	8/9
SN 1553	Tan Skiver Collar Bag, 7″ – – – – –	5/6
SN 1553	Tan Skiver Collar Bag, 8″ – – – – –	6/–

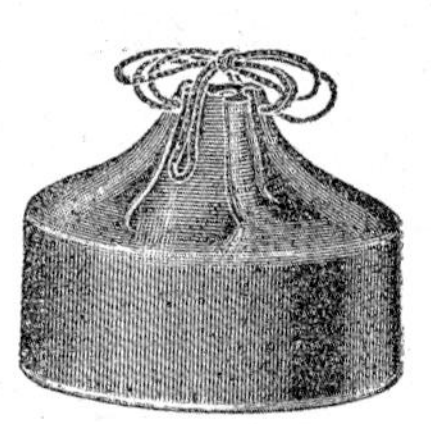

SN 1553

SHOOTING REQUISITES OF EVERY DESCRIPTION MADE TO ORDER ONLY

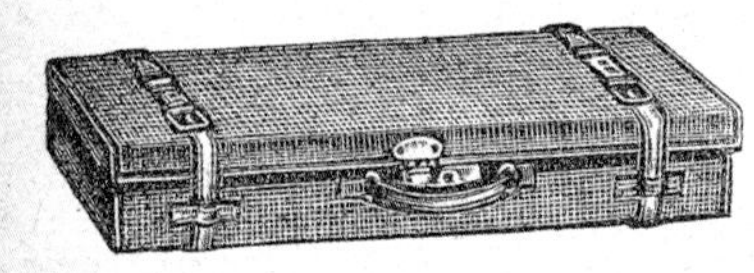

SN 1521

SN 1521 Gun Cases for 1 or 2 Guns, well fitted, lined good cloth, fitted with double-action lock, and straps all round. Brass corners extra.

Plain Leather Single 12/30 Gun Case – – – – –	56/– each
2nd Quality ditto – – – – –	42/– ,,
Best Green or Brown Canvas ditto – – – – –	43/– ,,
2nd Quality ditto – – – – –	38/– ,,
3rd Quality ditto – – – – –	29/6 ,,

SN 1522 Magazines fitted with movable partitions, double-action lock, brass corners and straps all round.

	500	400	300	200	
Superior Leather –	£5 5 0	£4 15 0	£4 5 0	£3 15 0	each
Plain Leather –	£4 4 0	£3 15 0	£3 3 0	£2 15 0	,,
Green or Mail Canvas	£3 5 0	£3 0 0	£2 10 0	£2 5 0	,,

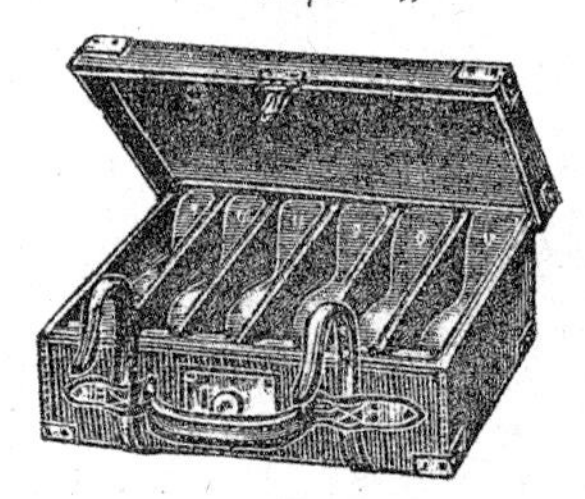

SN 1522

CARTRIDGE, GAME AND RABBIT BAGS ALSO MADE TO ORDER

All prices quoted in this List are subject to alteration without notice.

Edward Penton & Son

1–3 Mortimer Street
50 to 57 Newman Street, London N1
Northampton & Norwich
Factories: Wells Street, London W
Portland Street, Northampton

The registered trademark of Edward Penton & Son was the upper-case letter 'P' enclosed within a pentagon. This has not, however, been observed marked on their goods, though numerous products survive in good condition. The company distributed a hardback catalogue, at least throughout the 1920s. The range was wide, varying from buttonhooks through collarboxes, fibre and leather cases, to wardrobe and cabin-trunks. Many items now available originated from this prolific firm.

George Perry & Co.

Trunk Makers

11 Grafton Street, Dublin

Lovely, rare examples of leather and canvas-and-leather luggage from the later years of the 19th century. Often lined with blue-on-white striped ticking. The rectangular (with rounded corners), black-on-white, printed-paper label, applied to interior lid, adds the following: 'Bonnet Boxes, Warranted Solid Leather Portmanteaus, Travelling Bags, Hat Cases, Trunk and Rug Straps, American Trunks &c'.

Pescott

Maker

Leicester

Found impressed on simple, late 19th-century despatch-bag.

Opposite
1923–4 catalogue illustration showing kit bag, collarboxes and shooting cases by Edward Penton & Son.

G.G. Peterkin

Trunk & Bag Manufacturer

54 & 56 Schoolhill, Aberdeen

A red-and-black, oval, paper label to the linen lining of cowhide suitcases typical of the late 1920s and early 1930s, with heavy, cast-brass locks and leather cap corners deontes the products of this Aberdeen manufactuer.

Phillip Son & Nephew Ltd

Liverpool

This rare maker's name found gold-blocked onto Edwardian, black-morocco, motoring map-case (see illustration on p.16).

S.D. Piracha & Co.

26 Market Street, Bermondsey, London SE1

Reptile-skin merchant trading in the 1930s. It is interesting to note the cluster of such merchants in this area at this time, and to reflect that many items made from the skins they sold return to the famous antique market still held there on Friday mornings.

Pittway Bros.

8 & 9 Charles Street, Hatton Garden, London EC1

Advertised as dressing-case and dressing-bag makers in the early 1930s. No marked examples encountered.

'Pixie'

Brand name of lightweight luggage, most often found on medium grade vanity-cases *c.* 1950.

Pontings

'The Kensington Gift Store'

Kensington High Street, London W8

From their advertisement in a pre-Christmas newspaper, *c.* 1930, it is apparent that Pontings stocked a wide range of luggage. 'Ask for the Trunk Dept. (Basement)', it exhorts, and proudly illustrates items of luggage that ranged in price from 7/- to 69/6. These products were obviously aimed at the budget end of the market, and it is unlikely that many bore a Pontings retail label. It is interesting to note, however that many cases, strikingly similar to those in this advertisement survive, still in excellent condition and suitable for use, despite prices that were low even by the standards of the time.

Arthur Porter

4, Shepherd Street, Mayfair, London W1

Obscure trunk and portmanteau makers trading in the 1930s.

A. Potterton

47 Humberstone Gate, Leicester

Gold-blocked, rectangular, red-morocco trade label on finest-quality cowhide luggage states additionally: 'Bag, trunk, motor and sample-case manufacturers'. Maker's details are presented on each lateral side of the lid section, impressed directly onto the hide of the exterior – an attribute that hallmarks the item with the stamp of quality. Rare. Expect solid, cast nickel-on-brass locks, stamped 'secure lever'.

John Pound & Co. Ltd

1930s addresses:
81 & 82 Leadenhall Street, EC3
268 & 270 Oxford Street, W1
67 Piccadilly, W1
187 Regent Street, W1

Factories:
Imperial Buildings, Leman Street, E1
Late Victorian/Edwardian addresses:
81–84 Leadenhall Street
67 Piccadilly; 211 Regent Street
177, 178 Tottenham Court Road
378 Strand

Amongst the most hallowed of makers of the finest quality hide luggage, and even when found in poor condition, the uncompromising standard of excellence in manufacture that was adhered to at all times is still apparent. The company advertised variously as trunk and portmanteau makers ('also makers of compressed cane and fibre trunks') dressing-case makers, dressing-bag makers, bag makers, makers of motoring touring trunks and of 'travelling bags with toilet fittings complete'. Most examples will be of cowhide, often lined with morocco leather, but there was a wide variety over many years, and examples in fibre with linen lining survive, as do fine crocodile suitcases lined with silk. The maker's name, along with addresses current at time of manufacture, are usually gold-blocked either onto morocco linings or the hide body of the case. Heavy, cast nickel-on-brass locks and fittings, and usually fixed 'D' ring handle fixings.

W. Powell & Sons

35 Carr's Lane, Birmingham

Gunmaker and producer of superb hide cartridge-cases with brass-reinforced corners and locks.

Preedy

Makers

Manchester

Fairly prolific maker. Mid-Victorian examples bear the above details impressed on each lateral side of the lid section, whilst later examples may be marked with only an applied label to the interior lining.

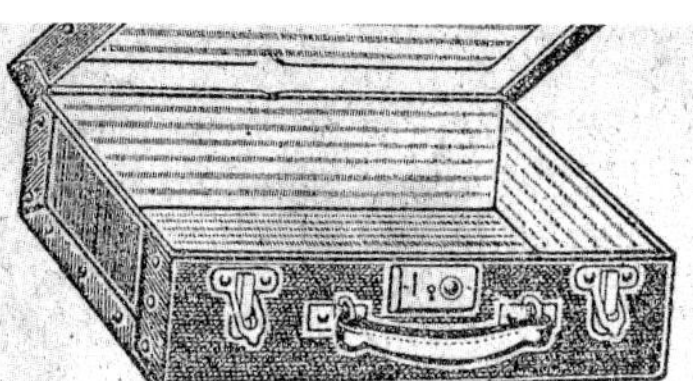

ASK FOR THE TRUNK DEPT. (*Basement*).

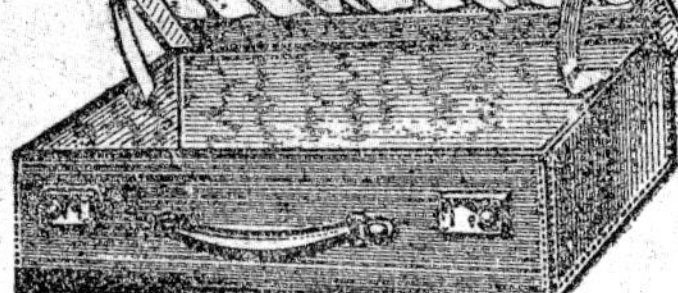

Exceptional Value.

LOT 7. Strong Attache Cases, Made on stout wood frames, covered smart dark Brown long grained leather cloth, good secure section lever lock and 2 securing clips, strong leather handle.

Sizes 14 x 9 x 4½ Special Value 7/-
,, 16 x 10 x 5 ,, 8/-
,, 18 x 11½ x 5½ ,, 9/-

To-day's Value 9/11, 10/11, 12/6.
16 x 18 in. have one lock and 2 clips, 14 in. one lock.

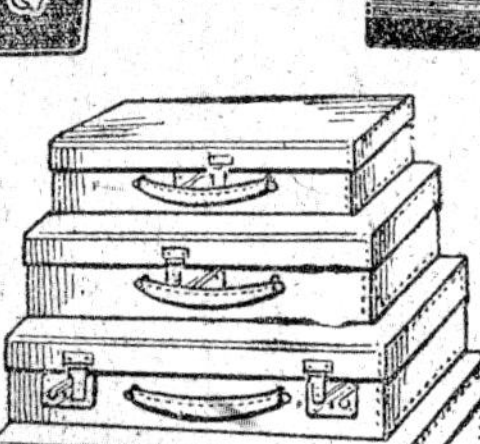

Extraordinary offer of a limited number of

WEEK END CASES

Made of Real Tan Cowhide lined moire, with expanding pocket in lid, hand sewn, 2 secure nickelled locks, strong handle, exceptional value. An ideal Yuletide present. Size 18 x 11 x 5.
Special Price **45/-**

LOT 1. Real Cowhide ATTACHE CASES. These cases are made of Dark Brown Cowhide, hand sewn throughout, smart lining, stout handle, 2 reliable nickelled lever locks, and two nickel patent receiving clips that make it impossible for case to open accidentally. To be cleared at prices that at least effect a saving of six shillings in the £.

Sizes and Sale Prices:

12 x 8 x 2¾ ... **19/11** | 14 x 9 x 3¼ ... **23/6**
16 x 10 x 3¾ ... **26/9** | 18 x 11 x 4 ... **29/6**

Postage 6d. extra.

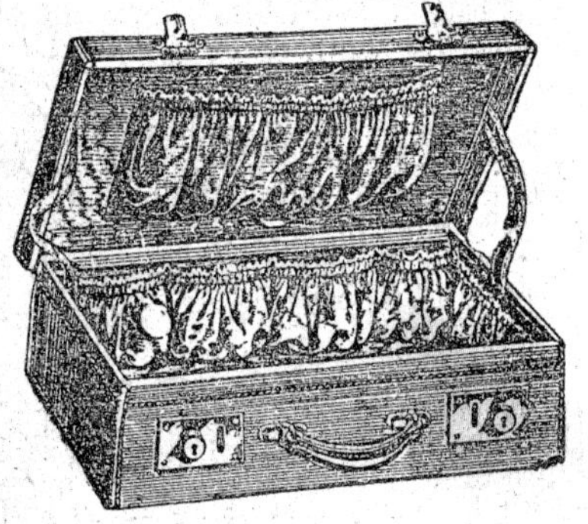

Two Most Useful Gifts.

Ladies' Unfitted DRESSING CASES, beautifully made in dark tan cowhide, hand sewn, light metal frame, lined good quality brown or blue moire with pockets all round and in lid, two secure nickelle locks. These cases are most practical and will take any size fittings.

Sizes 16 x 10 x 6 ... Special Price **55/-**
,, 18 x 11 x 5½ ... ,, **59/6**
,, 20 x 12 x 6 ... ,, **69/6**
,, 22 x 13½ x 6½ ... ,, **84/-**
,, 24 x 14 x 6½ ... ,, **94/-**

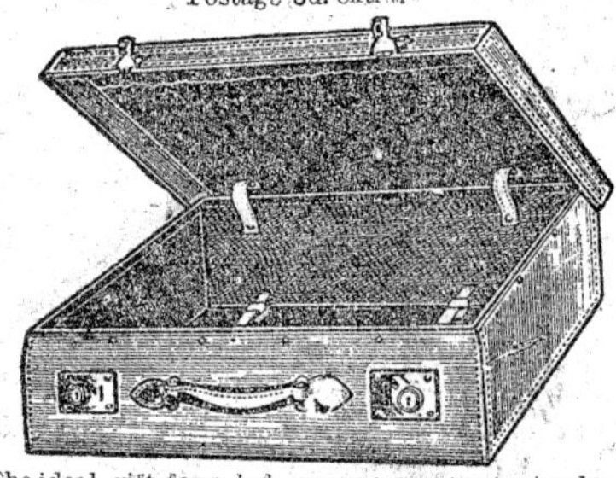

The ideal gift for a lady or gent. and the best value obtainable to-day, on each case there is at least a saving of 30/. Please compare our prices.
Specification as follows: Suit case made of high-grade Tan smooth cowhide, hand stitched, fitted with light weight metal frame, 2 solid brass double action lever locks, smart dark blue lining with expanding pocket in lid, 2 leather straps in bottom. strong handle, thoroughly well made and finished.

Sizes 18 x 13 x 6 ... Special Price. **49/6**
,, 20 x 13½ x 6 ... ,, ,, **55/-**
,, 22 x 14 x 6 ... ,, ,, **59/6**
,, 24 x 14½ x 6½ ... ,, ,, **65/-**
,, 26 x 15 x 6½ ... ,, ,, **69/6**

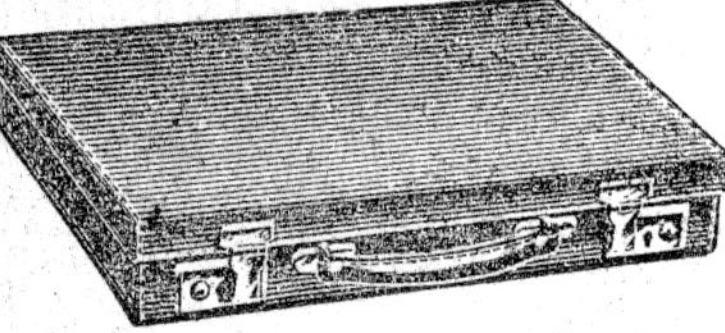

MUSIC ATTACHE CASES. Special Value, stout crocodile grained cowhide, 2 nickelled lever locks, also fitted with patent securing fasteners, hand sewn, a smart and useful gift. Size 15 x 11½ x 2½. Special Price **29/6**

Special Offer.
SOLID LEATHER MUSIC CASES made of stout grained cowhide thoroughly well made and beautifully finished, strong handle and polished metal bar.
Special Price **12/9**
To-day's value 20/-

STATIONERY ATTACHE CASES. Made of Real Tan Cowhide, hand sewn, fitted with expanding pockets for writing materials, address book, pen and pencil, and a supply of stationery. All fittings inside are covered with leather, 2 secure nickelled solid brass lever locks. A very smart and useful case. In two sizes only.

14 x 9½ x 3½. Special Price, **29/6**
To-day's value 45/-
16 x 10 x 3½. Special Price, **42/-**
To-day's value 55/-

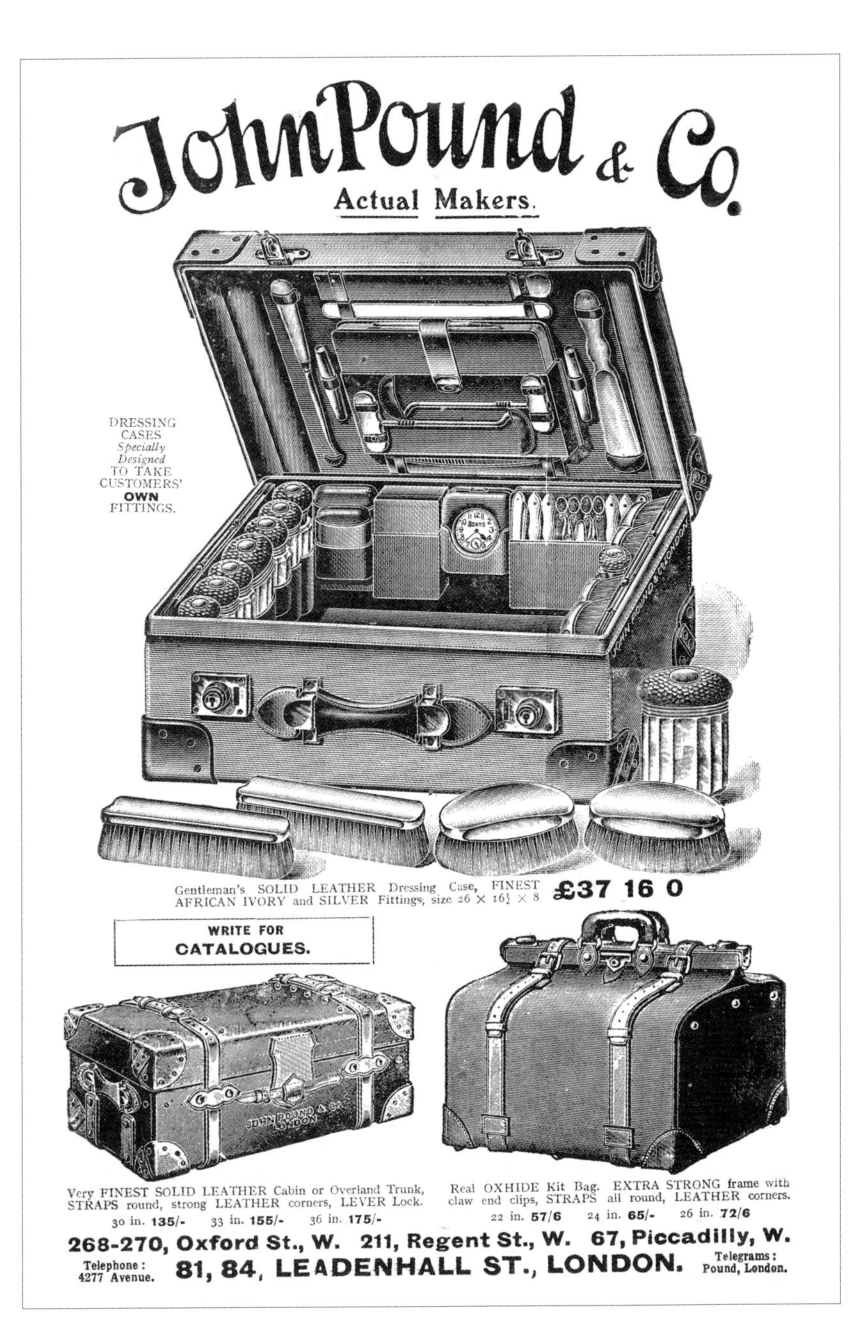

Opposite
Newspaper advertisement for Pontings Trunk Department *c.* 1935.

1913 advertisement for John Pound & Co.

Pukka Luggage

34 Bowling Green Lane, London

Brand name of one of the superlative types of vulcanised fibre luggage, produced mainly in the 1920s and 1930s. Products were sold with a 'guarantee bond' (see illustration on p.46), which reflects the pride that this firm took in its products; well deserved, since many items survive well beyond their envisioned useful life.

James Purdey & Sons Ltd

Current details:
57 South Audley Street, London W1
Tel: +44 (0) 171 499 1901

World-renowned and highly esteemed gunmakers. The substantial oak-lined gun and cartridge-cases bearing the label of this company are avidly sought and collected. James Purdey & Sons Ltd still manufacture these superb cases to the original quality and traditional specifications.

'Python'

Made in England

Brand name found on mid 20th-century, strengthened cardboard luggage, stamped silver-grey onto the front rim of the base with the logo of a coiled snake.

Railway Clearance Depot

Dealers

248 Vauxhall Bridge Road, London SW1

Advertised as leather case makers and trunk and portmanteau dealers in the mid 1930s.

Reid & Todd

8 Renfield Street, Glasgow
212 Sauciehall Street, Glasgow

Still famed as umbrella makers.

The following, taken from an advertisement of 1915 conveys the image of the company:

'Scotland's own specialist in TRAVELWARE
Two items which make our name famous North and South of the Tweed
The 'R & T' Umbrella @ 10/6
Crocus, Cherry, Pimento etc. Gold or Silver Mounts
Best Laventine Silk on Fox's Paragon frame.
'R & T' Suit Case @ 45/-
24-inch Fine Strong Hide
8 cap corners'.

S. Reid Ltd

90–91 Fleet Street, London EC4

Specialist manufacturers of shaped motoring luggage, mid 20th century. S. Reid Ltd also produced other types of luggage, notably some attractive satchel-style brief-cases of bridle hide.

Reptile Trading Co. Ltd

198 Bermondsey Street, London SE1

Reptile-skin merchant recorded in the 1930s.

Revelation Suit Case Co. Ltd

170 Piccadilly, London W1
Telegram Address: 'Ekspancase, Piccy'

The 'Revelation' expanding suitcase was a popular, very practical type of luggage that unfortunately ceased production around 1972, although a shop bearing this name still exists at the above address. The brand name was sold to Antler. It was constructed in two separate parts: a base and a lid section. These were joined together by four sliding, squared metal rods, within a similarly shaped casing that enabled the storage volume of the case almost to double in size, adjustable according to the volume of the contents. Often marked with the word 'Revelation' along with a black-stamped, imitation-ivory label bearing a logo of an angular bell-boy with striding legs and long arms echoing the expanding appendages of the case itself. The interior may also bear a paper label, an upright diamond, gold- or silver-foil-blocked onto black or blue, stating the brand name and relevant patents. Metal fixings were frequently stamped with the 'Revelation' logo.

W. Wood & Son Ltd advertised in the 1930s as 'Makers of the Revelation suit-case', and there was obviously widespread collaboration with various different manufacturers since, amongst others, the names of Finnigans, Lawrence & Sons and H. Boswell & Co. (Oxford) appear in partnership with that of 'Revelation' on examples of one of the most original and ingenious forms of the humble suitcase ever conceived. Many people covet such a suitcase – executed in almost every material imaginable. Apart from the usual cowhide, examples in cardboard and vulcanised-fibre exist, as do those in crocodile, and at least one in elephant hide. Revelation later produced striped, mottled-canvas luggage with smart cherry-red, bottle-green or navy-blue trim. This range, often with locks stamped 'Legge', included hatboxes, handbags, vanity- and shoe-cases in addition to standard suitcases.

1950s advertisement for S. Reid Ltd.

Reynolds

Hull

Mid 20th-century hide suitcases of fine traditional quality, but with less traditional brass locks typical of the 1940s onwards. 'Reynolds. Hull' impressed at the exterior, top-base section, above the handle, and 'Pat No 268857' discreetly stamped near the bottom-middle of each side base section.

J. & A.N. Richardson Ltd

49 & 51 Lisson Grove, London NW1

Recorded in the mid 1930s as suppliers of 'Picnic Sets (fitted)'.

David Richenberg

25 & 27 Clifton Street, London EC2

Suitcase maker of the 1920s/1930s.

E. Rickards

Maker

Bath

Good-quality cases by this maker are usually marked above the locks, rather than above the handle.

A. Rippengale & Co.

200 Hornsey Road, London N7

Attaché-case makers, mid 1930s.

Peter Robinson

Trunk & Bag Department

London

Oval, gold-blocked on tan leather, applied retail label with 'pinked' edge, or the more usual gold-stamped, red-morocco, oval label may be found on a variety of leather and leather-trimmed canvas luggage supplied in the early years of the 20th century by this well-known department store.

Paul Romand

213 Rue St Honore, Paris

Late 19th- to mid 20th-century maker of trunks and suitcases etc. of superb quality. Maker's name generally stamped onto heavy, cast-brass locks. (See illustration on p.69.)

J. W. Rose & Son Ltd

382 Brixton Road, SW9
161 Streatham High Road, SW16

Recorded as trunk and portmanteau makers in late 1920s/1930s.

Thomas E. Rowe & Co.

1 & 2 Gray's Inn Pass, WC1
26 & 27 Sandland Street, WC1

Advertised as trunk and portmanteau makers in the 1930s.

Ryan & Sons

Saddlers etc.

14–15 Spring Street, Hyde Park, London W1
Henley-On-Thames

Rare, classic, saddler-made suitcases. Early 20th century. Usually with a black-stamped, white-metal trade tag.

Saks Fifth Avenue

Fifth Avenue, New York, USA

Retailers for numerous years of a huge variety of fine luggage. Many vintage items of Louis Vuitton luggage also bear the retail label of this world-famous store, and certain companies such as the Hartmann Trunk Company supplied items exclusively for sale by Saks Fifth Avenue, including an imposing wardrobe-trunk *c.* 1915 labelled 'Pathfinder Imperial exclusively for Saks Fifth Avenue'.

Samsonite

Denver, Colorado, USA

Founded in 1910 by Jesse Shwayder, Samsonite is still thriving today. Probably one of the most widely recognised and sought-after brands of 1950s luggage in particular.

Montagu Arthur R. Saxby

22 & 24 Camden Road, NW1
19A Park Street, Camden Town, NW1

Advertised as trunk & portmanteau dealers in the 1930s. A company by the name of Montagu Saxby Ltd, Travel Goods, is currently trading at 8 Kentish Town Road, London NW1.

Scott's Hatters

1 Old Bond Street, London W1

Early-to-mid 20th-century leather gentlemen's hatboxes from this renowned hatmaker may bear the name and address directly impressed on the exterior-front lid of the case; alternatively, the details could appear on label applied to the interior.

G.W. Scott & Sons Ltd

1930s address:
144 Charing Cross Road, London WC2

Advertising in the early decades of the 20th century as makers of motor-car baskets, fitted picnic sets and tea baskets, often sold under the brand name 'Coracle' (see illustrations on pp.60) the case is generally marked 'Coracle' and the fittings either 'Coracle' or 'GWS & S' – sometimes a combination. Probably the most prolific producer of high-quality picnic sets in the 20th century.

Sam Segal

34 Paddington Street, London W1

Mid-1930s portmanteau dealer.

Selfridge & Co. Ltd

General stores

398–454 & 419–429 Oxford St, W1
1–4A, Orchard Street, W1
14–52 Duke Street, W1
2–13, 16, 22–45 Somerset Street, W1
101–107 & 125 Wigmore Street, W1
19–25 Irongate Wharf Road, Paddington, W2 London

This famed department store included an extensive luggage department, stocking a huge variety of luggage, as it does to this day. Examples of vintage luggage retailed by Selfridges in the earlier years of this century frequently bear an ivorene label announcing the name of the store.

Shuttleworth & Sons

Makers

Bridge Street, Chester

Late-Victorian makers of classic leather trunks and large suitcases. Red-morocco, trade label reads as above, along with the phrase 'Portmanteaus Trunks Bags Holdalls ETC'.

M.I. Siegenberg

131 & 133 Shaftesbury Avenue, London WC2

Recorded as trunk & portmanteau dealer, late 1920s/1930s.

'Sirram'

Picnic sets

Brand name of picnic sets sold by Marris's Ltd. (See above.)

Smith, Englefield & Co.

Bag & Trunk Makers

Nottingham

Products of this early 20th-century maker often bear a white-metal tag stamped black, giving the above details.

A. & J. Smith
Jewellers

Aberdeen

Circular, black-on-gold, transfer-applied retail label found next to gold-blocked 'Coracle' brand stamp to interior lining of six-person, wicker picnic set, *c.*1930.

George Smith & Co.
Manufacturers

151 Strand, WC London

Turn-of-the-century manufacturer of classic leather luggage, generally marked with the above details impressed in a straight line across the front-upper rim of the base section .

Stanley Griffin Smith
136 & 183 Praed Street, W2
70 Russell Square, WC1 London

Trunk and portmanteau makers in the mid 1930s. Various labels indicate that they traded under the name of 'Smith's Trunk Stores' and also had a branch at 85 Edgware Road, London W. This latter address appears to pre-date the Praed Street listing and the two have not been observed listed together. Smith's Trunk Stores sold a wide variety of luggage.

V. C. Smith
Maker

Shrewsbury

Early 20th-century maker whose name is often found impressed on the upper-exterior base section, above the handle.

W.H. Smith
Bath

Found inscribed on the interior metal rim of (usually) large, late-Victorian, leather kit-bags. Fairly rare.

Edward Hitchin Smyth
166A Upper Street, London N1

Advertised as trunk and portmanteau makers in the 1930s.

Soutar, Laird & Co. Ltd
Ladybank works, Dundee

'Manufacturers of waterproof canvas, school satchels, market bags, workmen's bags, tool rolls &c', *c.*1930s. Items rarely marked. (See Arnold Farthrop – London Agent.)

'SPARTAN'
Brand name appearing on lightweight, mid 20th-century cardboard luggage.

John Spurrier
145 Melbourne Grove, East Dulwich, London SE11

Leather case makers, mid 1930s.

Robert Stafford
19 Wagner Street, London E15

Attaché-case makers (1930s).

'Standex'
Brand name of leather luggage, *c.*1930, distinctive for the 'pat. pending' feature of a luggage tag integral to the handle fixing (see illustration on p.69). The 'Standex' brand name also appears on more conventional luggage.

H.W. Stanley
197 Brompton Road, London SW3

Advertised as trunk and portmanteau makers in the mid 1930s.

W. R. Stanley
142 Finchley Road, Hampstead, NW3 London

A rarely found, pasted-in, paper label depicting an attractive, striped trunk stating: 'Trunks repaired or taken in exchange' suggests that W.R. Stanley regularly dealt in second-hand items.

Stanley's Ltd
Travel goods

Kettering

1930s paper retail label applied to typical cases of the period, probably manufactured by Garstin's, Penton's, Pearson's or similar companies.

Stockland, Bennett & Co.
89 Tottenham Court Road, London W1

Trunk and portmanteau makers in the early decades of the 20th century.

Stockland, Tillett & Co.
6 Vernon Place, Bloomsbury Square, London WC1

Trunk and portmanteau makers, 1930s.

Miss Dorothy Stockland
3 Sicilian Avenue, London WC1

Advertised as trunk and portmanteau maker in the mid 1930s.

E. Strugnell
264 & 265 Upper Street, London N1

Advertised as attaché-case, leather case, trunk and portmanteau makers in the early decades of the 20th century. Also makers of 'featherweight' leather suitcases.

Suit Case Manufacturing Co. Ltd

See Attaché Case Manufacturing Co. Ltd

J. Sutton

George Street, Plymouth, England

J. Sutton applied an oval, gold-blocked, morocco-leather label to the interior lid of the suitcases they sold to indicate the point of purchase.

Swaine Adeney Brigg & Sons Ltd

1948 details: 185 Piccadilly, London W1
Telegraphic Address: 'Swaydeneyne, Piccy'
Current details:
10a Old Bond Street, London W1
Tel: +44 (0) 171 409 7277

Royal Warrant Holders as Whip, Glove & Umbrella Makers by 1948. Also produced fine, leather-cased accessories and wonderful, now very rare, luggage. Particularly sought after are the small cases holding a flask and sandwich tin.

Swaine Adeney Brigg still have a factory near Cambridge, and are Royal Warrant holders as suppliers of Leather Goods to H.M. the Prince of Wales, Whip & Glove makers to H.M. the Queen and Umbrella Makers to H.M. the Queen Mother.

The Swayne Adeney Brigg Factory produces leather goods for sale in the Old Bond Street shop and other highly regarded outlets.

Swaine & Adeney Ltd

1913 details: 185 Piccadilly, London W1
TN Regent 4277 (2 lines)

'Whip stick, umbrella & fine leather goods manufacturers; by appointment to HM the King & HRH the Prince of Wales'. (See above.)

W. Taborn & Co.

68 High Street, Tooting SW17
124 London Road, SE1

Trunk and portmanteau dealer, 1930s.

Tanner, Krolle & Co. Ltd

Current Head Office:
1A Mildmay Avenue, London N1
Tel: +44 (0) 171 359 0031
Shop: 38 Old Bond Street, London W1
Tel: +44 (0) 171 359 0031

Founded in 1865. Has consistently made classic leather luggage to the highest standard and quality, and continues to do so whilst also producing an exciting range of innovative modern handbags and leather goods, combining the virtues of traditional craftsmanship with the latest creative flair.

H. & W. Taylor

Makers

Baldwin Street, Bristol

Very obscure West Country leather suitcase maker. The few examples seen are identical in construction to cases produced by W. Insall & Sons, also of Bristol, even down to the unusual fact that the maker's name is stamped on each outer-lateral side of the lid section. Classic, dark-honey-coloured cowhide, and large, heavy cast-brass locks and fixings.

'Thermos' (1925) Ltd

Seymour Road, London E10
TN Leytonstone 4061–4

Manufacturers of 'Thermos' vacuum vessels, which by the late 1920s were replacing the earlier kettle, burner and wicker-covered flasks in many motoring picnic sets. Royal Warrant Holders by 1948, probably earlier.

W. Thomas & Son

71 Becklow Road, Shepherd's Bush, London W12

Trunk and portmanteau makers in the 1920s and 1930s.

W. Thornhill & Co.

144 & 145 New Bond Street, London

Highly regarded 19th- and early 20th-century maker of cowhide and crocodile leather goods, including many interesting early picnic sets, dressing-cases etc. By the 1930s the company was no longer listed. Exceptional quality, genuinely antique items made to the standards of the Victorian era.

Thresher & Glenny Ltd

152 & 153 Strand, London WC2

Early 20th-century trunk and portmanteau makers.

Tiffany & Co.

1930s addresses:
44 New Bond Street, W1 London
25 Rue de la Paix, Place de L'Opera, Paris
Fifth Avenue, 37th Street, New York.

A sought-after name for jewellery and lamps, Tiffany & Co. also created exquisite leather goods. The Paris branch closed during the Second World War and did not re-open, and the London branch, now at 25 Old Bond Street, does not retail leather goods. The Head Office at 5th Avenue and 57th Street in New York presides over more than 100 outlets worldwide, most of which still supply leather items. There are approximately twenty-five stores throughout the U.S.A., and branches in Australia, Canada, Germany, Italy, Japan, Korea, Singapore, Switzerland and Taiwan. (See illustration of crocodile bag on p.41.)

Moritz Tiller & Co.

Wien V11/2, Austria

Late 19th-century manufacturer of a variety of luggage, including black leathercloth, shaped hatboxes for a single top-hat (see illustration on p.70); red-lined, and wooden-hooped canvas suitcases with painted decoration.

Alfred Henry Tomkins

1 Gurney Road, Stratford, London E15

Leather case maker advertising in the mid 1930s.

Paul Tonnel

12 Rue de la Paix, Paris, France

Royal Warrant Holder in the second decade of the 20th century to the British Royal Family as Dressing Bag Maker.

'Travair'

Regd. Lightweight Luggage

Brand name found on mid 20th-century lightweight cardboard suitcases. Oval, paper trade label, silver-foil lettering on bottle green (see illustration on p.26).

Mark Utal

96 Queen's Road, London W2

Advertised as trunk and portmanteau maker in the 1930s.

Valijee & Sons.

Allibhoy, Multan Cantonment, Punjab

Royal Warrant Holders throughout the 20th century until at least 1948 as despatch-box manufacturers.

J.C. Vickery

1930s address:
143, 145 & 147 Regent Street, London

Exquisite-quality, much-collected antique leather goods, including suitcases and unusual items such as leather umbrella-cases. The address for the firm until the late 1920s was 179, 181 & 183 Regent Street.

In the 1930s J.C. Vickery advertised as goldsmiths, silversmiths, jeweller, dressing-case and high-class leather goods manufacturer to His Majesty the King & Her Majesty the Queen. Earlier it was phrased: 'Jeweller, Gold and Silversmith, & Travelling Bag Manufacturers. (Royal Warrant Holder)'.

'Victor'

Brand name of luggage produced by E.J. Pearson & Sons Ltd. Charming and collectable leather or canvas-and-leather luggage. Examples usually date from the mid 1930s onwards and combine classic features with slightly more modern styling.

Victoria Trunk Stores

31 Wilton Road, London SW1

Trunk and portmanteau dealers, early 20th century.

Louis Vuitton

Current details of four of around 150 branches throughout the world:
78 bis, Avenue Marceau, 75008 Paris
Tel: +33 (1) 47 20 47 00
54, Avenue Montaigne, 75008 Paris
Tel: +33 (1) 45 62 47 00
17–18 New Bond Street, London W1Y 9HF
Tel: +44 (0) 171 399 4050
198 Sloane Street, London SW1X
Tel: +44 (0) 171 235 3356

Arguably the most famous company ever to produce luggage. Whilst perhaps best known for its monogrammed canvas luggage covered with the yellow initials 'LV' and decorative floral motifs, the company has in fact, since 1854, produced a wide variety of every type of luggage imaginable, in virtually every possible material. Has been responsible for many innovations and trends, and its products have always been widely imitated. Maintains comprehensive archives and a museum of luggage and is ever helpful in matters of researching or repairing items of its luggage, whatever the age.

Most pieces bear a unique serial number on a label to the interior of the article, along with a lock or key number stamped on the patented lock, which is also stamped with the name and details. Every brass stud on a trunk or case will be stamped 'Louis Vuitton'. The traditional standards of service and quality are still upheld, and Louis Vuitton continue to design made-to-order pieces.

Maurice Waller

20A Newport Court, London WC2

Late 1920s/1930s trunk and portmanteau dealers.

George Henry Ward

71 Brixton Road, London SW9

Early 20th-century trunk and portmanteau makers.

Waring & Gillow Ltd

Oxford Street, London W1

Well known for furniture. Its catalogue of 'Oriental & Fancy Goods' indicates that the store had outlets in London, Manchester, Liverpool, Lancaster and Paris. In the 1930s the company sold a fine selection of luggage, from crocodile and cowhide dressing-, writing-, attaché- cases and suitcases to a variety of motor-cases and 'Combined Tea and Lunch' cases and baskets, which would now be referred to as 'picnic sets'.

Watajoy'

London

Attractive brand name, most often discovered on vellum and decorative canvas-covered cases. Some examples bear a white, red and blue woven label, 1 inch square.

Waterloo Trunk Stores

19 York Road, London SE1

Advertised as trunk and portmanteau dealers in the 1930s.

Magnificent Louis Vuitton monogram-canvas wardrobe-trunk, approximately 6 feet tall.

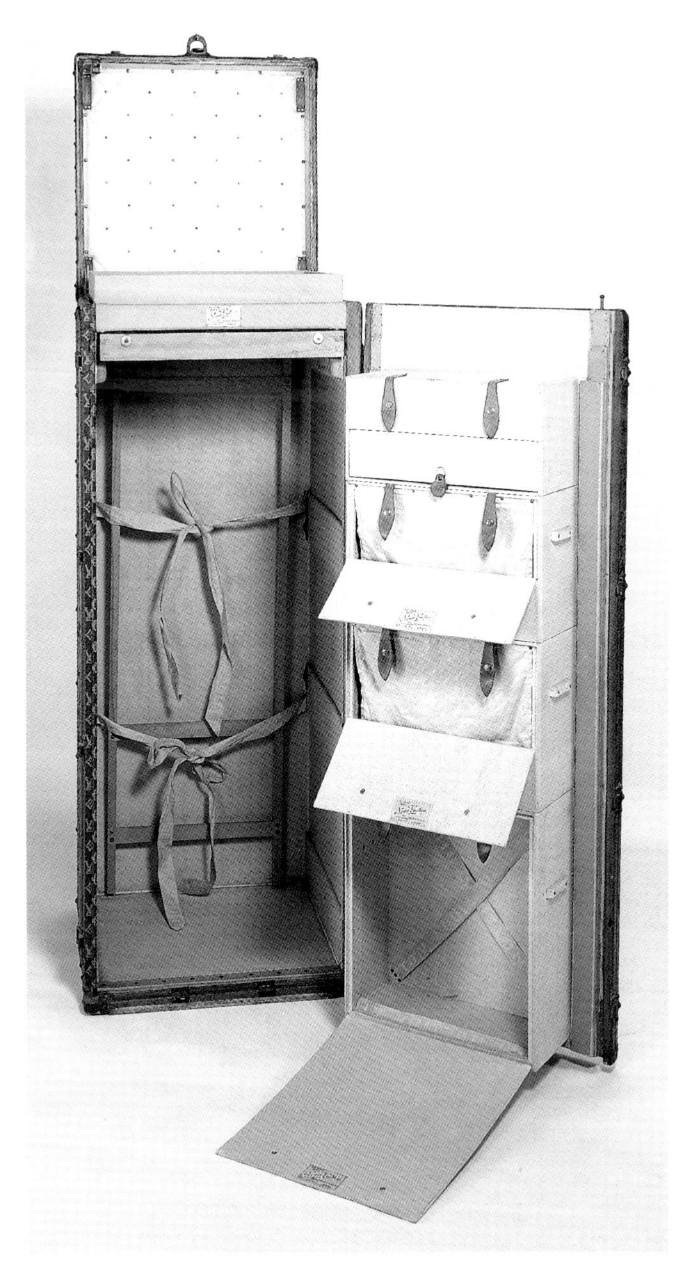

Watson Prickard

Liverpool

Classic, standard-sized hide suitcases, usually with leather cap corners and lined with black-morocco leather. Another interesting feature is the unusual construction of the handle: very flat with a rivet of copper or brass at each extremity.

Watt & Sons

Saddlers

Edinburgh, Scotland

Single, central, heavy cast-brass lock, and two straps with buckles either side of the lock are likely on cases made by this saddler.

Webb & Bryant

Makers

Queen Street, Portsea

Makers of various types of suitcases, trunks and hatboxes of cowhide in the 19th century. It is quite rare to find a maker's mark on a leather 'bucket-shaped' hatbox, but examples of single- and double-bucket hatboxes marked with this name on the base and front-inner rim have been observed in recent years. Also various rare examples of classic luggage with brass fittings, impressed with maker's name.

Webb & Son

Plymouth & Exeter

A gold-blocked, green-morocco label applied to the interior lid section of a rare Victorian suitcase with a distinctive handle fixing gave the above details. On later items the company favoured the use of an unusually shaped, applied label, gold-blocking the words 'Webb and Son Plymouth Ltd' in graduated letters exactly to fit the triangular snippet of toffee-coloured morocco.

Oliver Webb & Co.

15 Great Saffron Hill, London EC1

Recorded as dressing-case makers in the 1930s.

Webb's

9 Above Bar; 68 Oxford Street, Southampton

Early 20th-century makers of trunks, bags, cases and 'all travelling requisites'. A paper label applied to the interior of one trunk found was illustrated with a finely engraved row of six elephants' heads. Each holds a letter suspended from its proboscis, spelling out the word 'TRUNKS' (see illustration).

C.P. Weber

Stüttgart, Germany

Highly regarded manufacturer of leather travel goods. Closed at the end of the 19th century.

T. J. Weeks & Sons Ltd

152 Clarence Road, London E5

Advertised in the 1930s as attaché-case, leather case and dressing-case makers.

Western Trunk Stores

12 High Street, Notting Hill Gate, London W11

Listed in the 1930s as trunk, portmanteau and leather-bag makers.

'Wheary'

New York, USA

'Wheary' brand luggage is characteristic of the 1950s, and is all too infrequently encountered. A woven label, applied to the interior, reading 'WHEARY – the name to Remember – Racine New York' is typical. One stylish suite of three cases, graduated in size, is of bold and distinctive diagonally woven cream-and-brown canvas and has clear lucite handles with the brand name moulded integrally.

Whippy Steggall & Co

30 North Audley Street, London W1

Royal Warrant Holders in the second decade of the 20th century as Saddle and Harness Makers. Examples of specially commissioned luggage are exquisite and extremely rare. Stamped 'Whippy'.

William Whiteley Ltd

Queen's Road, London W2
33–41 Westbourne Grove, London W2
Porchester Gardens, London W2
Kensington Gardens Square, London W2

Advertising as 'universal providers' Whiteleys of Bayswater was, and still is, a renowned department store. Trunks and cases supplied by William Whiteley Ltd typically bore an imitation-ivory tag or a rectangular, morocco label, 2 x 3 inches,

Opposite
Page from Waring & Gillow catalogue.
Above
Detail from Webb's label.

FITTED CASES IN LEATHER AND BASKET-WORK

No. E33—Tea Case, Plywood, covered Rexine lined white Celaloid. Contains Aluminium Kettle with cork stoppers, lustre ware China, White China Provision Box, Wicker-covered Cream Flask, Tea, Sugar and Spirit Tins.
$18\frac{1}{2}$ x $9\frac{1}{4}$ x 8-in.—4 persons.
$22\frac{1}{2}$ x $10\frac{1}{2}$ x $8\frac{1}{2}$-in.—6 persons.

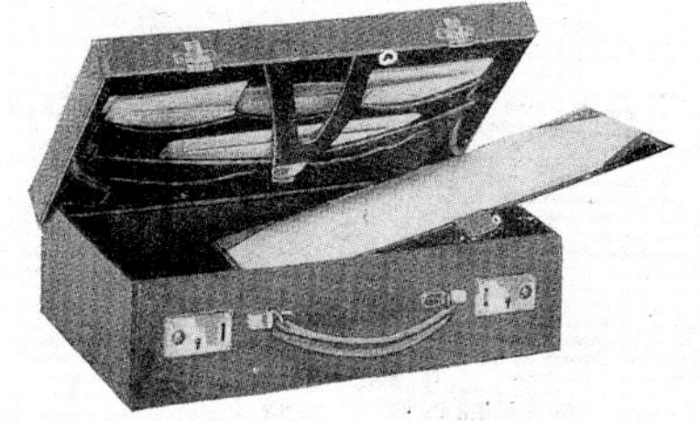

No: E39—Fitted Writing Attache Case in hide, lined leather, with best quality locks. Sizes 12, 13, 16 and 18-in.

No. E36—Tea Basket for 2 persons and 4 persons. Size $12\frac{1}{2}$ x $8\frac{1}{2}$ x $6\frac{1}{2}$ in.

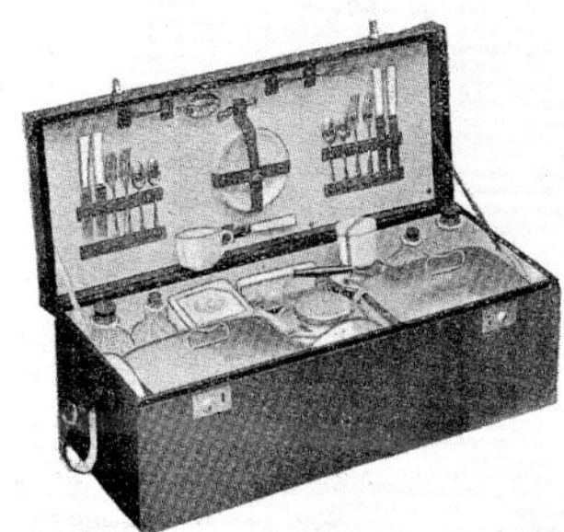

No. E34—Combined Tea and Lunch Case as 33 with additional fittings. Aluminium saucepan, Frying pan, 2 China Provision Boxes, Knives and Forks, Meat Dish, etc.
31 x 13 x 11-in.—4 persons

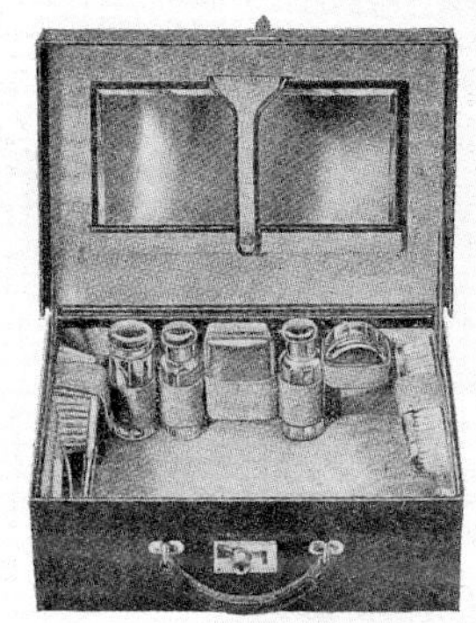

No: E40—Motor Case in Crushed Morocco, Gilt fittings and large standing Mirror, 13 x 9 x $4\frac{1}{2}$-in. Also in X grain.

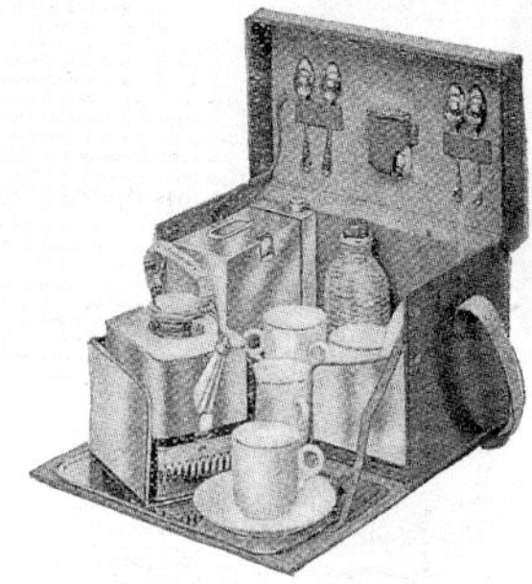

No. E37—Tea Case, solid Leather for 4 persons. Best Nickel Fittings and China Cups and Saucers. Size 12 x $7\frac{1}{2}$ x $7\frac{1}{2}$ in.

No. E35—Tea Case, Plywood, covered Rexine, filled with Patent Stand, Stove, Tea and Sugar Boxes, Spirit Tin, Kettle with screw lid to carry water, Tea Cups and Saucers, etc. 12 x 9 x $6\frac{1}{2}$-in.—2 persons.
13 x 10 x 7-in.—4 persons.

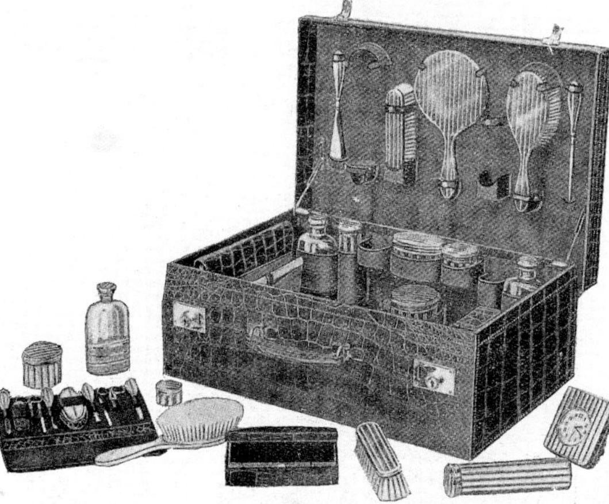

No. E41—Ladies' and Gent's Crocodile, X grain Morocco and Hide Dressing Cases, fitted to own requirements. Various qualities, sizes and designs in stock.

No. E38—Combined Lunch and Tea Basket for 2, 4 or 6 persons.
Size $16\frac{1}{2}$ x $12\frac{1}{2}$ x 8 in.—2 persons.
,, 19 x 14 x 8 in.—4 persons.
,, 19 x 14 x $8\frac{1}{2}$ in.—6 persons.

No. E43 — Real Crocodile Leather Hand Bags. Also in Ostrich and Lizard Skins, in various designs.

No. E42—Featherweight Leather Suit Case, superior finish, good locks. Sizes 20, 22, 24 and 26-in.
—Solid Leather Attache Case. Sizes 12, 14, 16 and 18-in.

LONDON MANCHESTER LIVERPOOL LANCASTER AND PARIS

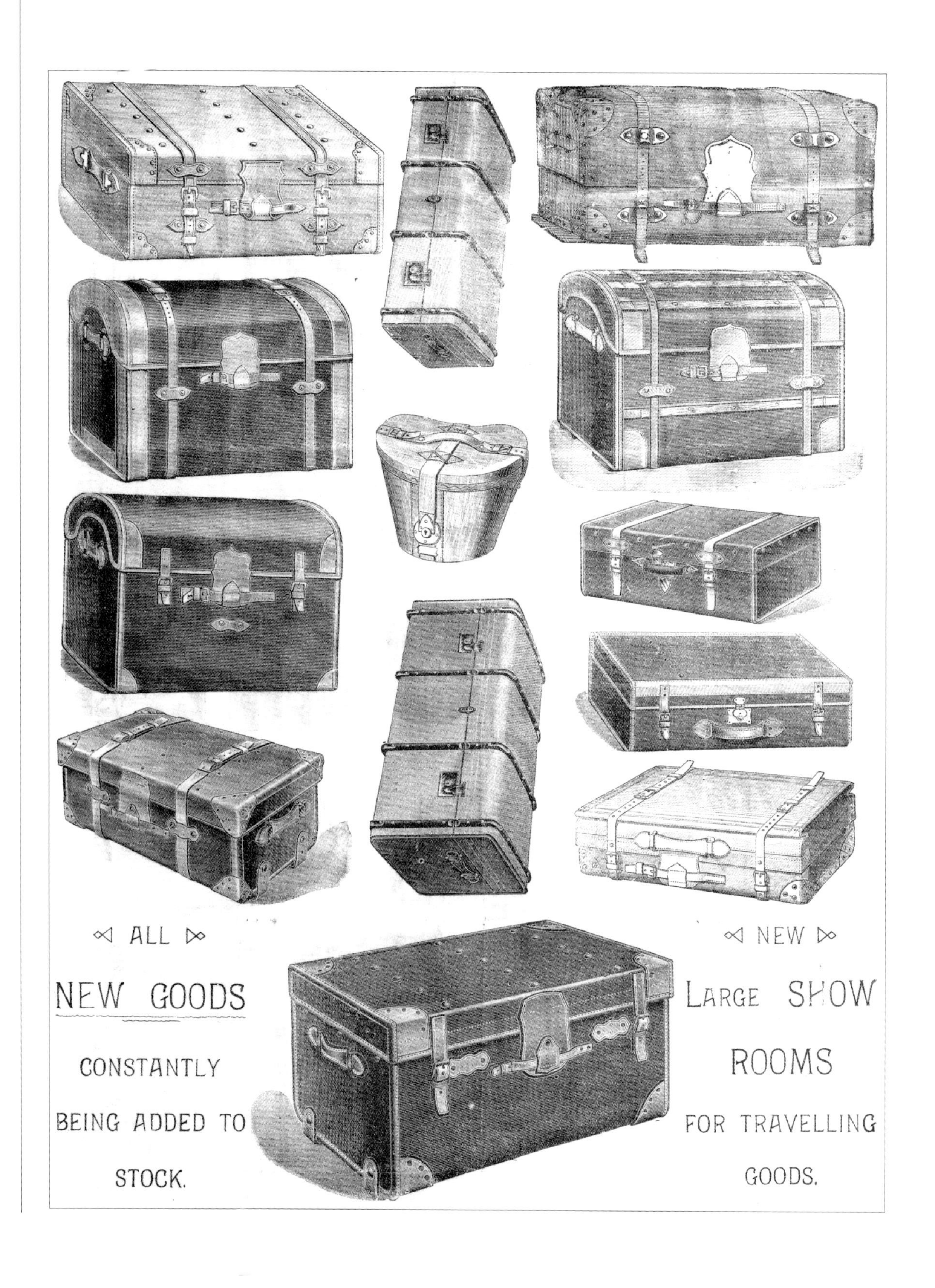
ALL
NEW GOODS
CONSTANTLY
BEING ADDED TO
STOCK.
NEW
LARGE SHOW
ROOMS
FOR TRAVELLING
GOODS.

99 *Opposite and above*
Late 19th-century flyer illustrating a variety of leather goods supplied by Charles J. Withnell & Sons of Scarborough.

gold-blocked with the Royal coat-of-arms and the words 'by special appointment to His Majesty the King William Whiteley Ltd Trunk Department 151 Queen's Road, Bayswater W'.

White's Stores

36 High Street Marylebone, London W1

Recorded as trunk and portmanteau dealer in the mid 1930s.

Whitmore's

New Street, Birmingham

The above details are invariably found stamped onto the nickel-on-brass locks of rare and exquisite dressing-cases by this maker, as well as being impressed onto the body of the case. Rare (yet possibly later) examples may bear a small, ivory tag stamped 'Whitmore & Son Ltd'.

Wickwar & Co.

6 Poland Street, London

Makers (amongst others) of government despatch-boxes, often black-morocco on a pine foundation. Items usually gold-blocked onto the interior rim with the above name and address, and the slogan 'Manufacturers to HM StatY Office' [sic].

W.W. Winship & Sons

USA

Early 20th-century maker specialising in iron-bound, vulcanised-fibre trunks.

Charles J. Withnell & Sons

4 Newboro' Street, Scarborough, England

Established 1859. Supplier of a huge variety of luggage and leather goods of superb quality via their 'The Leather Bag Warehouse' at above address. Also at St Nicholas Street, Scarborough.

S.B. Wolfsky & Co. Ltd

149 & 151 Pentonville Road, London N1

Advertised in the 1930s as dressing-case, dressing-bag and suitcase makers.

W. Wood & Son Ltd

Wholesale manufacturers

Kelvin Works, Power Road, London W4

Advertised as trunk, portmanteau and dressing-case makers. Also wholesale makers of leather bags and motoring touring trunks. In addition, this company were 'Makers of the Revelation suit case' the most attractive and widely renowned expanding suitcase ever produced. (See Revelation Suit Case Co. Ltd.)

Ernest Wright

97 Edgware Road, London W2

Advertised in the 1930s as a trunk and portmanteau dealer.

R.S. Wright & Son

122A Pentonville Road, London N1

Listed in the early 1930s as tea-basket maker.

M. Würzl & Sohne

Wien – Carlsbad, Austria

Occasionally found, usually on pigskin cases *c.*1920 with central lock and two side-catches similar to those standard on Louis Vuitton cases.

A Chronological History of Bramah Locks

DATE	CO. NAME/ADDRESS/EVENT	LOCK MARKINGS
21 August 1784	*Bramah Lock Company, Denmark Street, St Giles* *Lock Patent granted*	I (or J) Bramah. Patent
1784 2 June 1798	*Moved to 124 Piccadilly, London* *Patent Extended for a further 14 years*	(or J) Bramah with a crown I (or J) Bramah with a crown
1800		I (or J) Bramah with or without crown; 14 or 124 Piccadilly
1801	*Bramah Challenge Padlock, now in Science Museum, placed in shopfront of 124 Piccadilly*	
1805	*Prior to 1805 the 'bit' faced a smooth expanse of metal. After 1805 it was directly opposite one of the several slots*	
1813	*Eldest son, Timothy, joined business as a partner and company name changed to Bramah & Son*	
1814	*Joseph Bramah died*	
1821–1836	*Two other sons, Francis and Edward, became partners and company changed name to Bramah & Sons*	
1837–1841	*New partner joined company and name changed to Bramah and Robinson*	
1841	*Lock business separated from engineering business. Former became konwn as Bramah and Co., latter as Bramah, Prestige and Ball*	
1851	*Great Exhibition at Crystal Palace. Bramah & Co. still at 124 Piccadilly, London*	
1874–1901	*Company, now owned by J.T. Needs & Co., held Royal Warrants from Queen Victoria and King Edward VII. Works were in Deering Street. Company bought by Whitfield Safe & Lock Company, 1901*	J.T. Needs & Co. 100 New Bond Street (Late J Bramah 124 Piccadilly)
1910	*Businesses name changed back to Bramah & Co.* *Engineering works moved to Oldbury Place* *Royal Warrant granted by King George V*	
1926–1934	*Company took over 2 Nottingham Street*	2 Nottingham Street
1934	*Incorporated as Bramah Manufacturing Co. Ltd* *Name changed to Bramah's Ltd*	
1936–1939	*Moved from 2 Nottingham Street to 11 Old Bond Street*	11 Old Bond Street
1939 to 1963	*All bespoke locks made at Oldbury Place* *Company purchased by Mr Len Young, 1963, who within three months designed and launched the current low-volume, high-security range of products*	31 Oldbury Place
1966	*Bramah's Ltd purchased by J.R. Bramah and Co. Ltd and renamed to its current title: Bramah Security Equipment Ltd*	

Useful Contacts for the Collector

Kathy Adams Interiors
1509 Preston Road, Piano, Texas 75093, USA
Tel: +1 972 447 9231
Fax 1 972 447 0631

Antique City II
Lippenslaan 123, 8300 Knokke, Belgium
Tel + 32 (0) 5 062 3017
32 (0) 7 548 5060
Contact: Mike Vandenberghe

The Antique Travelshop
Versteeghstraat 67, Bussum, Netherlands
Tel: + 31 35 6933235
Fax: + 31 (0) 1969 624281
Postal address: P.O. Box 5815, NL-1410 GA Naarden, Netherlands

Sean Arnold Sporting Antiques
Viscount Court, 1 Pembridge Villas, London W2
Tel: + 44 (0) 171 221 2267/221 5464

Arundel Antiques Centre
51 High Street, Arundel, West Sussex, BN18 9AJ
Tel: + 44 (0) 1903 882749

ATGO Antiques
13–15 High Street, Owston Ferry, Near Doncaster, Yorkshire
Tel: + 44 (0) 142 772 8460
(By appointment)

Simon Baker Casemakers
Strathdon, Aberdeenshire, AB36 8US
Tel/Fax: + 44 (0) 19756 51777
http//:www.simonbaker.com
(By appointment)
Victorian-quality, hand-stitched, bridle-leather luggage made to order by Member of the Guild of Master Craftsmen. Simon Baker is a Director of the newly established Museum of Leathercraft
(http//:www.museumofleathercraft.com)

Beacon Hill Antiques
Jahnstrasse 60, (Nähe Eschersheimer Turm), 60318 Frankfurt Am Main, Germany
Tel: + 49 (0) 69 597 1561
Contact: Karen J. Nurdin

Barbara Becker
(Antique Luggage)
Hamburg
Germany
Tel: + 49 (0) 40 341 783

Bentleys
(Antique Luggage)
190 Walton Street, London SW3
Tel: + 44 (0) 171 584 7770
Contact: Tim Bent

Bonham's Chelsea Galleries
(Auctioneers)
65–69 Lots Road, London SW10
Tel: + 44 (0) 171 393 3900

Hans Broers, Carla Beek
(Picnic Set Specialists)
Insingerstraat 64, 3766 MC Soest, Netherlands
Tel: + 31 (0) 35 602 3032/694 2732
(By appointment)

Brooks
(Auctioneers)
81 Westside, Clapham Common, London SW4 9AY
Tel: + 44 (0) 171 228 8000
Fax: + 44 (0) 171 585 0830

Wilma Lee Brown Reasor Antiques
601 Belle Meade Boulevard, Nashville, Tennessee 37205, USA
Tel/Fax: + 1 615 353 0187
(By appointment)

Marie-France Bur
(Antique Luggage)
20, Avenue de Circourt
78170 La Celle St Cloud, France
Tel: + 33 1 39 18 34 37
Fax + 33 1 30 82 43 34
(By appointment)

Also at Old England store, Paris.

Carpoint
(Louis Vuitton Luggage)
Brauerstraße 41-43
D-76137 Karlsruhe, Germany
Tel: + 49 (0) 721 982120
Fax: + 49 (0) 721 9821226
Contact: Andreas Freund

Carr & Day & Martin Ltd
Lloyds House, Alderley Road, Wilmslow, Cheshire SK9 1QT
Tel: + 44 (0) 1625 539135
Fax: + 44 (0) 1625 548488

Manufacturers of 'Preslea' ('preserve leather') dressing, superb when used with caution on very dry leather. Also produces a range of metal polishes etc.

Chameleon
(Buyers & Suppliers of Nostalgia)
36 Church Street, London, NW8
Tel: + 44 (0) 171 723 9337

Christie's South Kensington
(Auctioneers)
85 Old Brompton Road, London SW7
Tel: + 44 (0) 171 581 7611

C.I.T.E.S. Management Authority France
Direction de la protection de la nature, Division de la Convention de Washington, 14 Boulevard du General-Leclerc, F-92524 Neuilly-sur-Seine Cedex, France
Tel: + 33 1 40812122 (Secretariat)
+ 33 1 40818447 (Importations)
+ 33 1 40818449 (Exportations)
+ 33 1 40818446 (Chef de Bureau)

C.I.T.E.S. Management Authority Germany
Bundesministerium für Umwelt, Naturshcutz und Readtorsicherheit, Naturschutz- Referat N1 3, Godesberger Allee 90, Postfach 12 06 29, D-53048 Bonn, Germany
Tel: + 49 228 305263034

The Bundesministerium does not issue C.I.T.E.S. permits or certificates.

C.I.T.E.S. Management Authority United Kingdom
Department of the Environment, Wildlife Licensing & Registration Branch, Room 822, Tollgate House, Houlton Street, Bristol BS2 9DJ
Tel: + 44 (0) 117 987 8749 (Enquiry Desk)/ + 44 (0) 117 987 8170
e-mail: global.wildlife@gtnet.gov.uk

C.I.T.E.S. Management Authority USA
US Fish and Wildlife Service, Office of Management Authority, 4401 N. Fairfax Drive, Room 420 C, Arlington, VA 22203, USA
Tel: + 1 703 3582093/ 1 703 3582095 (Chief & Operations Branch)
+ 1 703 3582104 (Branch of Permits)
e-mail: susan_lieberman@mail.fws.gov
or
kenneth_stansell@mail.fws.gov

Mary Clowney Antiques & Interiors
1009 Gervais Street, Columbia, South Carolina 29201, USA
Tel: + 1 803 765 1280
Fax: + 1 803 779 3060

Wilfried & Birgit Comberg Antiques
Luisenweg 25, 2300 Kiel 1, Germany
Tel: + 49 (0) 431 561800
+ 49 (0) 431 561 835
(By appointment)

Connolly Leather Ltd
Wandle Bank, Wimbledon, London SW19 1DW
Tel: + 44 (0) 181 542 5251
Fax: + 44 (0) 181 543 7455

Mail-order sales of 'Connolly Leather-Care' kits and 'Connolly Hide Food', an excellent treatment that will preserve and maintain leather. Always test treatments on an inconspicuous part of the item and carefully follow the directions on the packaging. Connolly Leather Ltd are well-known tanners, curriers and dressers of leather.

Connolly Leather Ltd
(Luxury Luggage and Accessories)
Shop:
32 Grosvenor Crescent Mews, London SW1
Tel: + 44 (0) 171 235 3883

The Connolly shop sells its leather-care kits together with its own exclusive range of avant-garde leather goods.

Chris Davey Antiques
Georgian Village
Camden Passage
London N1
Mobile: + 44 (0) 831 830540

Otten von Emmerich
(Louis Vuitton Luggage)
Bleichenhof – Passage, Bleichenbrücke 9, 20354 Hamburg, Germany
Tel/Fax + 49 (0) 403 571 3210
and
5191–3 Sabal Gardens Ln, Boca Raton, Fl. 33487, USA
Tel/Fax: + 1 561 989 0433
(By appointment)

Empire Luggage
Portobello Road, London W11
Tel: + 44 (0) 181 675 2069
Contact: Mia Cartwright

Finesse Fine Art
(Picnic Sets and Luggage)
9 Coniston Crescent, Weymouth, Dorset DT3 5HA
Tel: + 44 (0) 1305 770463
http://www.fine-art.demon.co.uk
(By appointment)

Gargoyles
(Decorative Antiques)
West 25th Street, New York, NY, USA
Tel: + 1 212 255 1035

Gliptone Leathercare U.K.
Manchester, M3 4NN
Tel: + 44 (0) 161 832 8532

Manufacturers of Gliptone Liquid Leather Conditioner: 'Restores, softens and protects with the advantage of restoring the smell.' Highly recommended, gentle leather conditioner.

Golden Oldies
(Decorative Antiques)
132–139 33rd Avenue, Flushing, New York, NY 11354, USA
Tel: + 1 718 445 4400
(By appointment)
Contact: Mark Weinstein

Goodison & Paraskeva Antiques
16 Camden Passage, Islington, London N1
Tel: + 44 (0) 0171 226 2423

John Gray
(Antique Luggage)
82b Portobello Road, London W11
Tel: + 44 (0) 171 229 2544

Greaves & Thomas
(Facsimiles and Reproductions)
Catherine Wheel Road, Brentford, Middlesex, TW8 8BD
Tel/Fax: + 44 (0) 181 568 5885
(By appointment)

Henry Gregory
(Antiques)
82 Portobello Road, London W11
Tel: + 44 (0) 0171 792 9221

Guinevere Antiques Ltd
578 King's Road, London SW6
Tel: + 44 (0) 0171 736 2917

John Gulshan Antiques
Cerney Mews, London W2 3DE
Tel: + 44 (0) 171 262 6434
Fax: + 44 (0) 171 262 6433
Mobile: + 44 (0) 802 754 622
(By appointment)

Guy's Luggage
Burton-on-Trent
Tel: + 44 (0) 1283 512036
(By appointment)

Hamilton White
(Picnic Sets and Luggage)
Littleheath Farm, Littleheath Lane, Lickey End, Bromsgrove, Worcestershire, B60 1HY
Tel + 44 (0) 1527 870633
Fax: + 44 (0) 1527 879817
Mobile: + 44 (0) 973 840668

Hastings Antique Centre
59–61 Norman Road, St Leonards-on-Sea, East Sussex TN38 0EG
Tel: + 44 (0) 1424 428561

The Hatstop
Camden Stables Market
Chalk Farm Road, London NW1
Tel: + 44 (0) 973 445537
Contact: Brian

Hide & Chic
(Antique Leather, Crocodile and Sporting Goods)
Tel: + 44 (0) 1454 273563
(By appointment)

Holland & Holland Ltd
(Antiques)
31 Bruton Street, London W1
Tel: + 44 (0) 171 499 4411

Adrian Hornsey Limited
(Antiques)
Three Bridge Mill, Twyford, near Buckingham, Buckinghamshire, MK18 4DY
Tel: + 44 (0) 1296 738373
Fax: + 44 (0) 1296 738322

De Jong Antiek
Showroom: Trade Mart Utrecht
Stand Number 4E.557
Utrecht, Netherlands
Tel/Fax: + 31 (0) 40 257 2606
Mobile: + 31 (0) 653 304883
Contact: Willem de Jong

Junktion
(Dealers in Nostalgia)
The Old Railway Station, New Bolingbroke, Near Boston, Lincolnshire
Tel: + 44 (0) 1205 480068 / 480087

M. & K. Ten Kate Antiques
Nes 89, Amsterdam, Netherlands
Tel: + 31 (0) 20 622 8133
(By appointment)

Dagmar Kölnberger
(Antique Luggage)
Gut Hausen, D-52072 Aachen, Germany
Tel: + 49 (0) 241 13271
Fax: + 49 (0) 241 175255
(By appointment)

John Lamdon Sporting Antiques
Warboys, Cambridgeshire
Mobile: + 44 (0) 831 274774
(By appointment)

Leatherbound
(Books & Antiques)
3350 Hardee Avenue, Atlanta, GA, 30341, USA
Tel: + 1 770 936 0528
Fax: + 1 770 936 0530
Contact: Jenny

McGowan & Rutherford Antiques Ltd
Crown House, Crown Street, Great Bardfield, Essex CM7 4SS
Tel: + 44 (0) 1371 810390
Fax: + 44 (0) 1371 810930
(By appointment)

Magus Antiques
187 Westbourne Grove, London W11
Tel: + 44 (0) 171 229 0267
Contact: Graham Walpole

Le Monde du Voyage
Stand 15- Marche Serpette, 93400 Saint-Ouen, France
Tel: + 33 40 12 64 03
Fax: + 33 44 84 00 36

Cosimo Morabito Antiques
Essen, Germany
Tel + 49 (0) 201 238492
Mobile + 49 (0) 171 805 4872

John Morgan
(Antique Luggage)
Mustons Yard, Shaftesbury, Dorset, SP7 8AD
Tel: + 44 (0) 1747 850353
Fax: + 44 (0) 1747 850354
(By appointment)

Peter Nelson Antiques
Preston, Lancashire
Tel: + 44 (0) 1772 721804
(By appointment)

Number 19
(Antique Luggage)
19 Camden Passage, Islington, London N1
Tel: + 44 (0) 171 226 1126
Fax: + 44 (0) 171 226 1991

Objects
(Antique Luggage)
9 Cork Street, 3rd Floor, Mayfair, London W1
Tel: + 44 (0) 171 792 3754
Postal address:
14 Denbigh Close, London W11
Tel: + 44 (0) 850 826 236
Fax: + 44 (0) 171 287 0365
Contact: T.S. Middlemas
(By appointment)

The Old Bag Company
Court Farm, Kington, Flyford Flavell, Worcestershire WR7 4DQ
Tel: + 44 (0) 1386 793427
(By appointment)
Contact: Lin Hopkins

Oxford
Country & Casuals
Osterstraße 30, 30159 Hanover, Germany
Tel: + 49 (0) 511 32 42 84
(By appointment)

Painted Cupboard Antiques
505 31st Street, P.O. Box 2927, Newport Beach, CA 92659, USA
Tel: + 1 714 723 4500
Fax: + 1 714 723 4501
(By appointment)
Contacts: Shannon & David C. Kerr

Pavilion
(Antique Luggage)
2 Styan Street, Fleetwood, Lancashire FY7 6SU
Tel: + 44 (0) 1253 779951
Fax: + 44 (0) 1253 779192
(By appointment)

Pullman Gallery
(Louis Vuitton and Hermés Luggage)
14 King Street
London SW1Y 6QU
Tel: + 44 (0) 171 930 9595

Raffles Antiques
40 Church Street, London NW8
Tel: + 44 (0) 171 724 6384

Reiseantiquitäten
(Travel-related Antiques)
S-Bahnbogen 200/Stand 2, im Antikmarkt Friedrichstrasse, 10117 Berlin-Mitte, Germany
Tel: + 49 30 208 26 81
Contact: Priti Shambhu

Relic Antiques
21a Camden Passage, Islington, London N1
Tel: + 44 (0) 0171 359 2597

Risky Business
(Antique Luggage)
44 Church Street, London NW8
Tel: + 44 (0) 171 724 2194
Contacts: Philip John & Chris Dobson

Sams & Son
(Antique Luggage)
Fraunhofferstrasse 23, D-80469 Munich, Germany
Tel: + 49 (0) 89 201 5735
Warehouse: Cornelinstrasse 19, D-80469 Munich, Germany
Tel: + 49 (0) 89 201 1152
Contact: Pierre Schmock

Sandra & Daniel Antiques
Portobello Road
London W11
Tel: + 44 (0) 171 229 8187
(By appointment)
Contact: C.J. Carter

Manfred Schotten
Sporting Antiques
The Crypt, High Street, Burford
Oxfordshire
Tel : + 44 (0) 199 382

Patrick en Martine Schroven
(Antique Luggage)
Slameuterstraat 7, B-2861
O.-L.-V.-Waver, Belgium
Tel: + 32 (0) 15 75.61.82
(By appointment)

Andrew Secombe
(Antiques)
High Street, Winchcombe,
Gloucestershire, England
Tel: + 44 (0) 124 260 3281

Lyn Sherer
Santa Monica Antique Market
1607 Lincoln Blvd, Santa Monica,
California 90404, USA
Tel: + 1 310 314 4894

C.W. Smith Inc.
(Antiques)
4424 Excelsior Boulevard, St Louis Park,
Minnesota 55416-4814, USA
Tel: + 1 612 922 8542
Fax: + 1 612 476 0883
(By appointment)

Sotheby's
(Auctioneers)
34–35 New Bond Street, London W1
Tel: + 44 (0) 171 493 8080
Fax: + 44 (0) 171 408 5989
US Headquarters:
1334 York Avenue, New York, NY
10021, USA
Tel: + 1 212 606 7000
Fax: + 1 212 606 7107
http://www.sothebys.com

Spencer Swaffer Antiques
30 High Street, Arundel, West Sussex
Tel: + 44 (0) 1903 882132

Sport is Life
(Luggage and Sporting Antiques)
Reading, Berkshire
Tel + (44) 734 751069
(By appointment)
Contact: Ron & Eileen

STV Hire, Props, Antique Hire
3 Ariel Way, Wood Lane, London W12
Tel: + 44 (0) 181 749 3445
Warehouse hiring out antique luggage

Jonathan Swire Antiques
Lytham St Annes, Lancashire FY8 1DA
Tel: + 44 (0) 1253 730076
(By appointment)

Julian Tatham-Losh Antiques
Brereton House, Stow Road,
Andoversford, Cheltenham,
Gloucestershire, GL54 4JN
Tel/Fax: + 44 (0) 1242 820646

Tennants Auctioneers
The Auction Centre, Leyburn, North
Yorks, DL8 5SG
Tel: + 44 (0) 1969 623780
Fax: + 44 (0) 1969 624281

Twin Peaks Antiques
Sint-Jorispoort 37, 2000 Antwerp,
Belgium
Tel: + 32 (0) 3 231 0523
+ 32 (0) 3 326 7963
Contact: Pascal Vandecasteele

Wilson Peacock
(Auctioneers)
Auction Centre, 26 Newham Street,
Bedford MK40 3JR
Tel: + 44 (0) 1234 266366

Wood and Brass from the Past
(Antiques)
Portobello Road, London W11
Tel + 44 (0) 973 551543
(By appointment)
Contact: Dee

World Wildlife Fund – UK
Panda House, Weyside Park, Catteshall
Lane, Godalming, Surrey GU7 1XR
Tel: + 44 (0) 1483 426444

XS Baggage Antique Luggage
Antiquarius, 137–141 King's Road,
London, SW3
Tel: + 44 (0) 171 376 8781

Michael Young Antiques
22 The Mall, 359 Upper Street,
Islington, London N1
Tel: + 44 (0) 171 226 2225

Index

References **in bold** are to illustrations

Hand luggage of the early 1930s. The 'Popular soft lid' is not greatly in demand amongst current collectors of luggage.